T0314966

Ethics and Global Environmental Policy

Ethics and Global Environmental Policy

Cosmopolitan Conceptions of Climate Change

Edited by

Paul G. Harris

Chair Professor of Global and Environmental Studies, Hong Kong Institute of Education

Edward Elgar
Cheltenham, UK • Northampton, MA, USA

Published by
Edward Elgar Publishing Limited
The Lypiatts
15 Lansdown Road
Cheltenham
Glos GL50 2JA
UK

Edward Elgar Publishing, Inc.
William Pratt House
9 Dewey Court
Northampton
Massachusetts 01060
USA

A catalogue record for this book
is available from the British Library

Library of Congress Control Number: 2011925736

ISBN 978 0 85793 160 3

Typeset by Servis Filmsetting Ltd, Stockport, Cheshire
Printed and bound by MPG Books Group, UK

Contents

Contributors

Nigel Dower is Honorary Senior Lecturer in Philosophy at the University of Aberdeen, Scotland. He taught at Aberdeen for three decades, and he has been a visiting scholar in Zimbabwe, Iceland and the United States. His interests include exploring ethics in a globalized world through teaching, lectures, writing and consultancy. He is author of *World Ethics: The New Agenda* (Edinburgh University Press, 2007), which is in its second edition, and *An Introduction to Global Citizenship* (Edinburgh University Press, 2003). He is co-editor with John Williams of *Global Citizenship: A Critical Reader* (Edinburgh University Press, 2002).

Romain Felli is a Visiting Research Fellow in geography at the University of Manchester and holds a fellowship from the Swiss National Science Foundation. Trained as a geographer and as a political scientist, he works on the global ecological crisis, climate refugees and the philosophy of political ecology. He is author of *Les deux âmes de l'écologie: Une critique du développement durable* [*The two souls of ecology: A critique of sustainable development*] (L'Harmattan, 2008).

Philip S. Golub is a political scientist and a specialist in international relations. He teaches international relations and international political economy at the Institute for European Studies of Université Paris 8, where he is Senior Lecturer, and the American University of Paris, where he is Associate Professor. He is a widely published author and Contributing Editor of *Le Monde Diplomatique.* His historical sociological study, *Power, Profit and Prestige: A History of American Imperial Expansion*, was published by Pluto Press (2010).

Paul G. Harris is Chair Professor of Global and Environmental Studies at the Hong Kong Institute of Education. He has written more than 100 scholarly journal articles and book chapters on global environmental politics, policy and ethics. He is author or editor of 15 books, recently including (as author) *World Ethics and Climate Change* (Edinburgh University Press, 2010) and (as editor) *China's Responsibility for Climate Change* (Policy Press, 2011).

Michael W. Howard is Professor of Philosophy at the University of Maine. He is the author of *Self-Management and the Crisis of Socialism* (Rowman and Littlefield, 2000) and editor of *Socialism* (Humanity Books, 2001). He is the national coordinator for the US Basic Income Guarantee Network.

Jennifer Kent is a doctoral candidate in the Institute for Sustainable Futures at the University of Technology, Sydney. Her research aims to determine how notions of moral responsibility at individual and collective levels are understood and practiced within Australian nongovernmental organizations working on community-based climate change action. She holds a Master of Environmental Law degree and has worked for many years in the environment field, initially in the area of solid waste and more recently in sustainability education positions in state and local government and in the nongovernmental sector.

Jean-Paul Maréchal is Associate Professor of Economics at Université Rennes 2. He is a researcher at the Institute for Applied Mathematics and Economics and a member of PEKEA (political and ethical knowledge on economic activities). His research fields are economic ethics, environmental economics, sustainable development and epistemology. He has published five books, one of which was awarded the 2001 prize of the French Academy of Moral and Political Sciences, and numerous articles in scholarly journals and specialized publications.

Robert Paehlke is Professor Emeritus of Environmental and Resource Studies at Trent University in Peterborough, Ontario. He is a founding editor (1971–1981) of the journal *Alternatives: Canadian Environmental Ideas and Action.* His books include *Some Like It Cold: the Politics of Climate Change in Canada* (Between the Lines, 2008), *Democracy's Dilemma: Environment, Social Equity and the Global Economy* (MIT Press, 2004) and *Environmentalism and the Future of Progressive Politics* (Yale University Press, 1991). He is also co-editor of *Managing Leviathan: Environmental Politics and the Administrative State* (Broadview Press/ University of Toronto Press, 2005).

Steve Vanderheiden is Associate Professor of Political Science and Environmental Studies at the University of Colorado at Boulder, where he specializes in normative political theory and environmental politics. In addition to numerous journal articles and book chapters on a variety of topics in environmental political theory, he is author of *Atmospheric Justice: A Political Theory of Climate Change* (Oxford University Press,

2008), which won the 2009 Harold and Margaret Sprout Award from the International Studies Association's Environmental Studies Section, and editor of *Political Theory and Global Climate Change* (MIT Press, 2008).

Preface

Climate change is the greatest challenge facing humanity. Yet efforts by the world's governments to restrain the pollution that causes it have failed utterly. This may appear to be a harsh assessment given the great amount of work that has been done by diplomats, policymakers and officials over the last three decades and more. It may seem to ignore the innovations and policy changes that have occurred, often in response to initiatives by national governments and the international community. But everything that is happening to address climate change and its causes – and increasingly its consequences – is found wanting when we realize that global warming is continuing apace. Indeed, greenhouse gas pollution, particularly pollution emanating from the burning of coal and other fossil fuels, is *increasing* quite rapidly due to continued emissions from the rich world and accelerating emissions from the developing world – not to mention other worrying trends that will make the problem far worse in the future, such as the release from melting permafrost of methane, a very powerful greenhouse gas, which in turn leads to more warming, more melting and more methane in a potentially devastating 'positive feedback' loop. The feeble results of international negotiations among parties to the United Nations Framework Convention on Climate Change, as evidenced by diplomatic conferences held in Copenhagen in 2009 and Cancun in 2010, as well as national policies that do far too little to cut greenhouse gases substantially, show that governments are not yet up to the task of responding effectively to climate change. Put another way, all efforts by governments to address global warming, while welcome and worthy of recognition, are simply far too little relative to the scale of the problem.

What is one to think, and what is one to do, in the face of feeble policies and actions by governments? If the current driving forces behind international negotiations and policies on climate change – most prominently the self-perceived national interests of states and the narrow priorities of influential industries – are preventing effective action, is there a better basis for responding to the problem? Who or what can fill the breach of failed state-oriented policies? In an attempt to answer these and similar questions, I have joined others in looking for alternative ways of conceiving of climate change. In particular, I have sought to look at it through the lens of cosmopolitanism (most prominently in another book, *World Ethics and Climate Change* [Edinburgh University Press, 2010], which informs this one). From a cosmopolitan perspective, ethical obligations and responsibilities are not defined or delineated by national borders; human beings, rather than states, ought to be at the centre of moral calculations. This worldview points to climate-related policies and actions that are less 'international' and more 'global', thus encompassing actors other than just states, and specifically including individual human beings. A central aim of this book is to help establish some lessons for climate change policy that can come from conceiving of the problem from this perspective.

The idea for the book began as a panel on 'Cosmopolitan Diplomacy and the Climate Change Regime' at the 2009 general conference of the European Consortium for Political Research in Potsdam, Germany. Some of the participants in that panel have contributed to this volume, while others joined the project later. I am grateful to all of those who have been involved for sharing their thoughts and observations. Work in this book was substantially supported by a grant from the Research Grants Council of the Hong Kong Special Administrative Region, China (General Research Fund Project No. HKIEd 340309). I wish to acknowledge gratefully the comments of anonymous reviewers, which have helped to strengthen the whole book and individual chapters. My thanks go to the kind and capable people at Edward Elgar for taking on this project and bringing it to readers. As always, I am especially grateful to K.K. Chan for a decade of daily support that makes the long process of completing each new book less arduous.

The potential value of cosmopolitanism for developing new policies to combat climate change is that it helps us break free of the myopic obsession with states and their governments, in the process highlighting the role of people (and other non-state actors) as causes of – and solutions to – the problem, not to mention highlighting the role of millions of people as its victims. One of the messages emanating from this book is that capable individuals everywhere have an ethical obligation to help in the fight to mitigate the causes and consequences of climate change. With this in mind, all of the editor's royalties from the sale of the book will be paid to Oxfam directly by the publisher. This is a small gesture toward partially fulfilling this obligation. It is also a recognition of a new reality: if all of us who are polluting the Earth's atmosphere and have the capability to change our ways do not do so, there will be a boundless need to help those most affected by the inevitable hardships and suffering that climate change will bring.

Paul G. Harris
Hong Kong

1. Introduction: cosmopolitanism and climate change policy

Paul G. Harris

Climate change is the most profound environmental problem facing the world – and possibly the most important problem of any kind in the long term. The latest science of climate change shows that massive cuts in emissions of greenhouse gas emissions will be needed by mid-century to avert extreme, possibly catastrophic, harm to Earth's climate system. Yet, despite ongoing and sometimes intense diplomatic efforts over two decades, governments of the world have been unable to agree to anything near the kind of regulation of pollution that would be required to undertake these cuts.[1] This was amply demonstrated by the much-anticipated December 2009 international climate change conference in Copenhagen, which failed to reach any formal or binding agreement on steps to reduce greenhouse gas emissions or to deal with the impacts of global warming. The Copenhagen conference, and the subsequent conference of the parties in Cancun a year later, revealed what may be a fundamental flaw in the international management of climate change, namely underlying norms and ethics that give overriding importance to states and their national interests, rather than to the people and groups who ultimately cause and are most affected by climate change.

A major manifestation of this problem is recurring debate over the historical responsibility of developed states for climate pollution. While those countries surely deserve blame if we think only in terms of states, this focus on state responsibility fails to account for rising greenhouse gas emissions among affluent people in the historically less responsible countries of the developing world. Given this growing misfit between historical national responsibility and current emissions, the emphasis on states rather than people may have to be overcome if the world is to take the extraordinary steps necessary to combat climate change aggressively in coming decades.

One major step toward this objective may be to look at climate change from a cosmopolitan perspective. Cosmopolitanism points toward politically viable alternatives to the status quo regime that are just, practical and – most importantly – potentially more efficacious than existing responses to climate change. Cosmopolitan conceptions of who is to blame for climate change, and whose rights are most in need of protecting in this regard, may usefully supplement the statist approach to the problem so far. At the very least, cosmopolitanism conceptions of climate change help us to identify fundamental problems with existing responses to this problem. Indeed, if taken seriously, cosmopolitanism forces a reevaluation of the causes and consequences of climate change while offering constructive critiques of the status quo.

This collection of essays undertakes this cosmopolitan reevaluation of the world's responses to climate change as part of a larger effort to understand how ethics can inform environmental governance. The contributors' arguments and analyses draw upon philosophy and ethics to inform the politics and policy of climate change.

FEATURES OF COSMOPOLITANISM

In contrast to the state-centric norms that have guided and indeed defined the international system for centuries, cosmopolitans envision an alternative way of ordering the world.[2] Cosmopolitans want to 'disclose the ethical, cultural, and legal basis of political order in a world where political communities and states matter, but not only and exclusively'.[3] States matter greatly, to be sure. But this is more of a practical matter than an ethical one for cosmopolitans. Thomas Pogge sums up three core elements of cosmopolitanism this way:[4]

> First, *individualism*: the ultimate units of concern are *human beings*, or *persons* – rather than, say, family lines, tribes, ethnic, cultural, or religious communities, nations, or states. The latter may be units of concern only indirectly, in virtue of their individual members or citizens. Second, *universality*: the status of ultimate unit of concern attaches to *every* living human being *equally* – not merely to some sub-set, such as men, aristocrats, Aryans, whites, or Muslims. Third, *generality*: this special status has global force. Persons are ultimate units of concern for *everyone* – not only for their compatriots, fellow religionists, or such like.

David Held has synthesized cosmopolitanism into a set of eight, universally shared, key principles: '(1) equal worth and dignity; (2) active agency; (3) personal responsibility and accountability; (4) consent; (5) collective decision-making about public matters through voting procedures; (6) inclusiveness and solidarity; (7) avoidance of serious harm; and (8) sustainability'.[5] From these principles a 'cosmopolitan orientation' emerges: 'that each person is a subject of equal moral concern; that each person is capable of acting autonomously with respect to the range of choices before them; and that, in deciding how to act or which institutions to create, claims of each person affected should be taken equally into account'.[6] Importantly for climate change, the last two principles provide 'a framework for prioritizing urgent need and resource conservation. By distinguishing vital from non-vital needs, principle 7 creates an unambiguous starting point and guiding orientation for public decisions [and] clearly creates a moral framework for focusing public policy on those who are most vulnerable.'[7] A 'prudential orientation' is set down by principle 8 'to ensure that public policy is consistent with global ecological balances and that it does not destroy irreplaceable and non-substitutable resources'.[8]

Some cosmopolitans take a consequentialist perspective, such as Peter Singer's utilitarianism, while others take a deontological perspective, such as Simon Caney's global political theory premised on human rights.[9] Charles Jones describes three 'species' of cosmopolitanism: utilitarianism, human rights and Kantian ethics.[10] He defines cosmopolitanism as a moral standpoint that is 'impartial, universal, individualist, and egalitarian. The fundamental idea is that each person affected by an institutional arrangement should be given equal consideration. Individuals are the basic units of moral concern, and the interests of individuals should be taken into account by the adoption of an impartial standpoint for evaluation.'[11] The nature of cosmopolitanism might be best appreciated by pointing to what it rules out: 'it rules out the assigning of ultimate rather than derivative value to collective entities like nations or states, and it also rules out positions that attach no moral value to some people, or weights the value people have differently according to characteristics such as ethnicity, race, or nationality'.[12] Another way of looking at cosmopolitanism, particularly in practice, is that it 'does not privilege the interests of insiders over outsiders'.[13] In a fundamental way, what is crucial about the cosmopolitan perspective is its 'refusal to regard existing political structures as the source of ultimate value'.[14]

Two versions of cosmopolitanism are routinely identified: an ethical/moral/normative version, which focuses on the underlying moral arguments regarding how people, states and other actors should justify their actions in the world, and an institutional/legal/practical version, which aims to translate ethics into institutions and policies. Pogge distinguishes between moral and legal cosmopolitanism. Moral cosmopolitanism points to the moral relations among people; 'we are required to respect one another's status as ultimate units of moral concern – a requirement that imposes limits on our conduct and, in particular, on our efforts to construct institutional schemes'.[15] Legal cosmopolitanism goes a step further by advocating creating institutions of global order, possibly in the form of a 'universal republic' in which 'all persons have equivalent legal rights and duties'.[16] This latter position may seem to be a bit extreme; moral cosmopolitanism certainly does not require institutionalization of a universal republic (or 'world government'). One variant of institutional cosmopolitanism asserts that 'the world's political structure should be reshaped so that states and other political units are brought under the authority of supranational agencies of some kind'.[17] Institutional cosmopolitans sometimes call for major, even radical, changes to global institutions, but moral cosmopolitans frequently do not see this as being necessary.

An alternative (more realistic) version of institutional cosmopolitanism 'postulates fundamental principles of justice for an assessment of institutionalized global ground rules [while also being] compatible with a system of dispersed political sovereignty that falls short of a world state'.[18] As Caney points out, some moral cosmopolitans 'reject a world state. They think that cosmopolitan moral claims are compatible with, or even require, states or some alternative to global political institutions.'[19] Thus it is entirely possible and appropriate to advocate institutions well short of world government that contribute to global order generally, and particularly global justice within specific issue areas. What is more, as Darrell Moellendorf reminds us, 'very few people who have thought about these matters [i.e., whether an egalitarian world order would contain multiple states or a world-state] have considered the latter a real possibility, and with good reason'[20] – not least the practicality of governing the world's many billions of people and the threat such a world state might pose to human rights. However, it is also clear 'that the establishment and maintenance of justice requires a significant re-conceptualization of the principle of state sovereignty [and] a coordinated international response'.[21]

Cosmopolitanism includes two additional features according to Brock and Brighouse: identity and responsibility.[22] The former refers, for example, to a person who is influenced by a variety of cultures or perhaps one who identifies with broader humanity rather than to a particular group or nation. The latter 'guides the individual outwards from obvious, local, obligations, and prohibits those obligations from crowding out obligations to distant others . . . It highlights the obligations we have to those whom we do not know, and with whom we are not intimate, but whose lives touch ours sufficiently that what we do can affect them.'[23]According to Robin Attfield,

> Cosmopolitan ethicists maintain that ethical responsibilities apply everywhere and to all moral agents capable of shouldering them, and not only to members of one or another tradition or community, and that factors which provide reasons for action for any agent, whether individual or corporate, provide reasons for like action for any other agent who is similarly placed, whatever their community may be or believe. They also deny limits such as community

> boundaries to the scope of responsibilities; responsibilities (they hold) do not dwindle because of spatial or temporal distance, or in the absence of reasons transcending particular facts or identities.[24]

One might also think of both weak and strong forms of cosmopolitanism, the former saying that some obligations obtain beyond the society or the state, while the latter says that any principles (of justice, for example) that apply within the state also apply worldwide. As Brock and Brighouse see it, 'everyone has to be at least a weak cosmopolitan now if they are to maintain a defensible view, that is to say, it is hard to see how one can reject a view that all societies have *some* global responsibilities'.[25]

Pogge addresses critics of 'weak' cosmopolitanism – 'the anodyne view that all human beings are of equal worth', which almost everyone, except 'a few racists and other bigots', accepts – and 'strong' cosmopolitanism – 'the view that all human agents ought to treat all others equally and, in particular, have no more, or less, reason to help any one needy person than any other', which it might be argued is falsely expansive – by proposing an 'intermediate' view of cosmopolitanism based on *negative* duties.[26] From this viewpoint, the fact that someone is a fellow national citizen 'makes no difference to our most important negative duties':[27] 'You do not have more moral reason not to murder a compatriot than you have not to murder a foreigner. And you do not moderate your condemnation of a rapist when you learn that his victim was not his compatriot.'[28] Intermediate cosmopolitanism 'asserts the fundamental negative duty of justice as one that every human being owes to every other'.[29] But just as duties of justice vary *within* communities – it is widely accepted that one can have a greater duty to family members than to the wider community – this does not mean that there are no duties whatsoever, in particular that there is no duty to avoid contributing to conditions that undermine the fundamental rights and needs of others within the community. Similarly, while we may favor compatriots in many ways, we ought not to support institutions that impose an unjust order on people living in other communities. According to Pogge, 'special relationships can *increase* what we owe our associates, but cannot *decrease* what we owe everyone else'.[30] The upshot is that, 'though we owe foreigners less than compatriots, we owe them something. We owe them negative duties, undiluted.'[31]

For cosmopolitans, 'the world is one domain in which there are some universal values and global responsibilities'.[32] Cosmopolitan responsibility entails 'the recognition that since we live, in some sense, in one global community or society – whether or not most of us have much of a feeling for this – we do have duties to care in one way or another about what happens elsewhere in the world and to take action where appropriate'.[33] It is not enough to identify with humanity to be a cosmopolitan; it is necessary to act (or be willing to act) accordingly. From this basis, it stands to reason that capable individuals are obliged to act even if they live in dissimilar communities (that is, rich or poor countries), and those who are more capable are more responsible to do so. James Garvey puts it this way: 'the better placed an individual is to do what is right, the greater the onus on him to do what is right.'[34]

Cosmopolitans frequently justify their claim that justice ought to prevail globally using one or both of two arguments. One argument, sometimes building on John Rawls's domestic theory of justice, is that levels of international cooperation today are extensive enough to make international society sufficiently like domestic society to warrant applying justice principles that were previously the domain of domestic communities to world affairs.[35] Another argument, derived from the empirical realities of globalization and the interdependencies and cause-and-effect relationships it manifests, is that justice ought to prevail globally because people and communities, whether knowingly or not, intentionally or not, increasingly affect one another, sometimes in profound ways. Justice is demanded by this latter argument because globalization is in large part a process of redistribution of scarce resources away from those with the least to those with the most. David Weinstock describes a relatively new 'way of understanding the relationship between the global rich and the global poor[:] the fate of the global rich is not as causally independent of the plight of the global poor as had previously been thought . . . According to this view, globalization makes it the case that our obligations toward the global poor are obligations of *justice* rather than of *charity*. . .'.[36] Climate change could be the most profound manifestation of this latter argument.

COSMOPOLITAN JUSTICE

Most cosmopolitans accept, and often advocate, duties of *global* justice for states and frequently by individuals. Global justice is based upon a cosmopolitan world ethic premised on the rights, duties and ethical importance – and moral pre-eminence – of persons. According to Dower, the wish for global justice is motivated by three claims: (1) 'obligations are substantial or significant, rather than minimal or merely "charity"'; (2) global obligations should be premised on 'institutional arrangements which specify quite clearly which bodies have which duties to deliver justice'; and (3) obligations have their foundation in 'the human rights of others which are either violated by the global economic system or fail to be realized because of it'.[37] For cosmopolitans, 'the world is a community of people and not a set of countries: that is, it is a community in which all have a claim to justice, just as they themselves owe justice to others'.[38]

Onora O'Neill proposes a practical approach to determining who has moral standing: 'Questions about standing can be posed as context-specific *practical* questions, rather than as demands for comprehensive theoretical demarcations.'[39] Answers are found in part in the assumption that people 'are already building into our action, habits, practices and institutions'.[40] This suggests a 'more or less cosmopolitan' approach to principles of justice in given contexts.[41] O'Neill's practical approach offers a *relational* account of moral standing:

> Conjoined with the commonplace facts of action-at-a-distance in our present social world, this relational view points us to a *contingently* more or less cosmopolitan account of the proper scope of moral concern in some contexts. We assume that others are agents and subjects as soon as we act, or are involved in practices, or adopt policies or establish institutions in which we rely on assumptions about other's capacities to act and to experience and suffer. Today we constantly assume that countless others who are strange and distant can produce and consume, trade and negotiate, . . . pollute and or protect the environment . . . Hence, *if* we owe justice (or other forms of moral concern) to all whose capacities to act, experience and suffer we take for granted in acting, we will owe it to strangers as well as to familiars, and to distant strangers as well as to those who are near at hand . . . Today only those few who genuinely live the hermit life can consistently view the scope of moral concern which they must acknowledge in acting as anything but broad, and in some contexts more or less cosmopolitan.[42]

This is a view of justice that takes *obligation* as being essential; 'obligations provide the more coherent and more comprehensive starting point for thinking about . . . the requirements of justice' than do rights because it is hard to know who has harmed which distant others.[43]

Andrew Dobson makes a case for cosmopolitan obligation arising from the causal impacts of globalization in its many manifestations, including global environmental change.[44] What is especially important about his argument is that he goes beyond cosmopolitan morality and sentiment, which are important but apparently not sufficient to push enough people to act. Dobson describes 'thick cosmopolitanism', in particular the source of obligation for cosmopolitanism, in an attempt to identify what will motivate people (and other actors) not only to accept cosmopolitanism but to act accordingly. While he seems to accept that we are all members of a common humanity, he is unhappy with leaving things there: 'Recognizing the similarity in others of a common humanity might be enough to undergird the principles of cosmopolitanism, to get us to "be" cosmopolitans (principles), but it doesn't seem to be enough to motivate us to "be" cosmopolitan (political action).'[45] Common humanity is one basis for cosmopolitanism, but it does not create the 'thick' ties between people that arise from causal responsibility. Dobson argues that the way to think about the 'motivational problem is in terms of nearness and distance . . . to overcome the "tyranny of distance"'.[46] He invokes Linklater's suggestion that if we are causally responsible for harming other people, and the physical environment upon which they rely, we are far more likely to act as cosmopolitans should.[47] Relationships of causal responsibility 'trigger stronger senses of obligation than higher-level ethical appeals can do'.[48]

To help us comprehend this connection between nearness, causality and motivation, Dobson describes a Good Samaritan whose actions to assist a suffering man move us because the Samaritan was not responsible for the man's injuries; the Good Samaritan acted purely out of beneficence. However, if the Samaritan were 'implicated in the man's suffering in one way or other, we would *expect* him to go to his aid and his act of succor would seem less remarkable'.[49] This illustrates the 'cosmopolitan nearness' that arises from causal responsibility.[50] While this might not be the whole story – we might have obligations to help others in critical need simply because we are *capable* of helping them – even if we do have obligations for other reasons they are amplified if we are indeed the cause of the harm in question. What is more, in keeping with cosmopolitan morality, the 'causal responsibility approach' that Dobson describes is universal: 'the obligation to do justice implicit in it is owed, in principle, to absolutely everyone without fear or favor'.[51] It is also universal because we now live in a globalized world in which most of what we, the affluent people of the world, do involves relations of causal responsibility, therefore making them relations of justice. As Dobson puts it, 'the ties that bind are not, therefore, best conceived in terms of the thin skein of common humanity, but of chains of cause and effect that prompt obligations of justice rather than sympathy, pity, or beneficence'.[52] O'Neill argues that 'in our world, action and inaction at a distance are possible. Huge numbers of distant strangers may be benefitted or harmed, even sustained or destroyed, by our action, and especially by our institutionally embodied action, or inaction – as we may be by theirs.'[53] Dobson's point is that in these kinds of relations of actual harm, justice 'is a more binding and less paternalistic source and form of obligation than charity'.[54]

The cosmopolitan standpoint presents serious challenges to the prevailing climate change regime, but it also offers opportunities for new, possibly more effective prescriptions for making the regime more attuned to reality and thus for making it more effective. What more do cosmopolitans say about climate change? The next section scratches the surface of their arguments, setting the stage for further discussion in the chapters that follow.[55]

CLIMATE CHANGE AND COSMOPOLITANISM

Lorraine Elliott asserts that environmental harms crossing borders 'extend the bounds of those with whom we are connected, against whom we might claim rights and to whom we owe obligations within the moral community'.[56] She describes this as a 'cosmopolitan morality of distance', which effectively creates 'a cosmopolitan community of duties as well as rights'.[57] Elliott argues that this obtains for two reasons: 'the lives of "others-beyond-borders" are shaped without their participation and consent [and] environmental harm deterritorialises (or at least transnationalises) the cosmopolitan community. In environmental terms, the bio-physical complexities of the planetary ecosystems inscribe it as a global commons of a public good, constituting humanity as an ecological community of fate.'[58] Consequently, Elliott believes that the cosmopolitan standpoint provides a better 'theoretical and ethical road map for dealing with global environmental injustice' than does international doctrine.[59] Attfield goes further, arguing that only cosmopolitanism can do 'justice to the objective importance of all agents heeding ethical reasons, insofar as they have scope for choice and control over their actions, and working towards a just and sustainable world society'.[60] He believes that criticisms of failed state responses to environmental problems will inevitably be based on cosmopolitanism because 'the selective ethics of nation states are liable to prioritize some territories, environments, and ecosystems over others. If this meant nothing but leaving the other environments alone, this might not be too pernicious. [However,] it often means not leaving alone the others but polluting or degrading them.'[61] Derek Heater also critiques what he calls the 'traditional linear model of the individual having a political relationship with the world at large only via his state' because, at least if we are concerned about 'the integrity of all planetary life, the institution of the state is relegated to relative insignificance – if not, indeed, viewed as a harmful device'.[62]

This points to the need for a theory of environmental justice that fully encompasses the causes and consequences of climate change. Such a theory almost certainly must be cosmopolitan, as Steve Vanderheiden argues:

> Insofar as a justice community develops around issues on which peoples are interdependent and so must find defensible means of allocating scarce goods, global climate change presents a case in which the various arguments against cosmopolitan justice cease to apply. All depend on a stable climate for their well-being, all are potentially affected by the actions or policies of others, and none can fully opt out of the cooperative scheme, even if they eschew its necessary limits on action. Climate change mitigation therefore becomes an issue of cosmopolitan justice by its very nature as an essential public good. . . .[63]

Governments have agreed to some principles and practices of environmental justice that apply at the *interstate* level, including in the context of the climate change regime. Indeed, some of the related proposals have a cosmopolitan flavor. For example, the developing countries have called for the allocation of greenhouse gas emissions to be based on equal per capita allotments. This would in effect require the rich countries to pay poor ones for the use of the latter's allotments. While the climate change negotiations so far have arrived at bargains that fall short of codifying these equal per capita rights to the atmosphere, Frank Biermann believes that only such allotments 'have an inherent appeal due to their link to basic human rights of populations in both South and North' and will probably have 'the normative power to grant the climate governance system the institutional stability it needs in the decades and centuries to come', not least because only equal per capita rights can be democratically supported.[64] Peter Baer also argues that equal per capita emissions rights are the only ethical option, noting that this has the practical benefit of offering options for developing country emissions limits in the future.[65] But others have argued that 'political and economic reasons [mean that] such a proposal has no chance of being accepted by developed countries because it leads to unacceptable costs for them. . .'.[66] This skepticism is well justified based on the history of the climate change regime, even as these arguments show that cosmopolitan-like positions are being debated among diplomats already – although states are often the intended bearers of duties and frequently the proposed beneficiaries of associated rights.

The work of Simon Caney is particularly noteworthy for the way it looks at climate change from a cosmopolitan standpoint, in particular showing how and why climate change is unjust because it threatens human rights. As Caney states, 'the current consumption of fossil fuels is unjust because it generates outcomes in which people's fundamental interests are unprotected and, as such, undermines certain key rights . . . This is unjust whether those whose interests are unprotected are fellow citizens or foreigners and whether they are currently alive or are as yet not alive.'[67] His argument proceeds this way: (1) persons have a right to something if it is 'weighty enough to generate duties on others'; (2) climate change jeopardizes 'fundamental interests' (for example, not suffering from drought, crop failure, heatstroke, infectious diseases, flooding, enforced relocation and 'rapid, unpredictable and dramatic changes to their natural, social and economic world'[68]); (3) the interests jeopardized are of sufficient weight to generate obligations on other persons; thus (4) 'persons have a right not to suffer from the ill-effects associated with global climate change'.[69] One advantage of Caney's argument is that it does not turn on the question of who is causing climate change.

But the question still remains: who ought to bear the burdens of addressing the problem? Caney answers that all persons 'are under the duty not to emit greenhouse gases in excess of their quota' and persons 'who exceed their quota (and/or have exceeded it since 1990) have a duty to compensate others (through mitigation or adaptation)'.[70] He concludes that 'the most advantaged have a duty either to reduce their greenhouse gas emissions in proportion to the harm resulting from [mitigation] or to address the ill-effects of climate change resulting from [adaptation] (an ability to pay principle)', with the added proviso that 'the most advantaged have a duty to construct institutions that discourage future non-compliance'.[71]

Michael Mason argues that there is environmental responsibility across borders because those who produce significant harm, regardless of whether they are states, are morally obliged to consider those affected by the harm, regardless of whether those harmed are co-nationals.[72] What he argues for requires 'an appreciation of expressions of well-being not mediated by states'.[73] Jamieson has pointed out that the notion that governments have duties only to one another is problematic for environmental protection.[74] Given the nature of environmental problems and the environmental interests and actions of different individuals and organizations, 'rather than thinking about the problem of the global environment as one that involves duties of justice that obtain between states, we should instead think of it as one that involves actions and responsibilities among individuals and institutions who are related in a variety of different ways'.[75] Consequently, the common notion of international environmental justice – obligations of states to aid one another in this context – ought to:

> be supplemented by a more inclusive ecological picture of duties and obligations – one that sees people all over the world in their roles as producers, consumers, knowledge-users, and so on, connected to each other in complex webs of relationships that are generally not mediated by governments. This picture of the moral world better represents the reality of our time in which people are no longer insulated from each other by space and time. Patterns of international trade, technology, and economic development have bound us into a single community, and our moral thinking needs to change to reflect these new realities.[76]

COSMOPOLITAN CONCEPTIONS OF CLIMATE CHANGE

What comes from these views is a need to interrogate the preoccupation with governments and states, and potentially to focus much more on the needs, obligations and actions of individuals if we are to find alternatives to the weak international climate change regime. Contributors to this book start from this need by exploring new ways for the world to respond to the crisis of climate change. To be sure, the contributors to this volume sometimes differ in their conceptions of cosmopolitanism, in keeping with the thinkers cited above. Nevertheless, they share Pogge's fundamental concepts – individualism, universality and generality – as starting points for conceptualizing climate change in a new way, and they agree that looking at climate change from a broadly cosmopolitan perspective can shed light on alternatives to the largely failed status quo. We begin with chapters focusing on what cosmopolitanism tells us about individual responsibilities for climate change before moving to discussions of justice among states and how cosmopolitanism might inform diplomacy and policies related to climate change.

In Chapter 2 Steve Vanderheiden argues that justice requires adequate action on climate change mitigation *and* adaptation. This raises the theoretical question of how national inadequacy in climate change policies affects the ongoing assignment of related burdens, something that is of tremendous policy relevance in ensuring that the normative objectives of global climate policy are achieved. Vanderheiden believes that the normative concept of responsibility offers a value basis for linking mitigation and adaptation efforts under a single overarching conception of justice, thereby providing a coherent account of climate justice. His view of climate justice links it to an account of responsibility, which can be stated briefly as demanding that persons and peoples voluntarily take responsibility for the climate change that they culpably cause, or be held responsible for it by others. If this can be done for all persons and peoples that affect or are affected by climate change, climate justice can be usefully understood as an effort at ensuring globalized responsibility. Being responsible in this sense requires that persons and peoples avoid harming others through the environmental externality of climate change, whether by paying the relevant mitigation costs needed to avoid causing climate change or by paying the adaptation costs need to avoid this resulting in human harm. Insofar as persons and peoples fail to do their share in mitigating or controlling this global environmental problem, they can be held responsible by others through assessments of liability to pay compensation.

In Chapter 3 Nigel Dower examines climate change from the perspective of selected cosmopolitan theories. From these theories he derives the cosmopolitan responsibility of individuals. As he points out, even if cosmopolitanism can be translated into practical climate policies, individuals will have to take some responsibility for bringing about the needed changes. Put another way, cosmopolitan responses to climate change need to occur at the level of institutions, including what Dower calls 'cosmopolitical changes' *and* at the level of 'active global citizenship engagement'.

In examining the latter, Dower focuses on three questions: if effective international cooperation to address climate change is to be realized, how important is it to allow for a variety of pragmatic principles, such as precaution, 'contraction and convergence' and 'polluter pays', and how significant are ethical principles that different individuals and groups can accept? What is the nature and extent of the obligations of individuals with respect to climate change, particularly those whose lifestyles are carbon-intensive, here and now – prior to any changes in laws, regulations, economic incentives or social expectations? And what is the relevance of these individual obligations for the likelihood and legitimacy of government policies for addressing climate change?

In Chapter 4 Jennifer Kent looks at individual responsibility and voluntary action on climate change. Consistent with Dower's argument, Kent believes that individuals and households will need to contribute to efforts to significantly reduce greenhouse gas emissions if international targets are to be achieved. However, she believes that the role of individual responsibility as a component of government policies, as well as of discourse around climate change, is under-theorized. Consequently, the mechanisms for ascertaining what individuals should do, and how local actions link to those at the global level, are poorly understood. She points out that responsibility for mitigating climate change has generally been applied through determining how greenhouse gas cuts can be distributed fairly among states. Much less emphasis has been placed on how these contributions will be distributed between states and their citizens, least of all within an ethical framework that establishes rights to, and responsibilities for, the global atmospheric commons. With this in mind, Kent's chapter considers individual responsibility for climate change mitigation as it is expressed through forms of voluntary action, and she considers how perceptions of agency may influence change at other levels. She shows that the adoption of a cosmopolitan ethic within the global climate regime is currently hampered by limits on individual agency.

Looking at the development of the international legal regime on climate change from the perspective of the capitalist relations of production, in Chapter 5 Romain Felli argues that developments in the regime are flawed due to three forms of 'fetishism': distribution, global governance and the state. Because of these so-called fetishisms, a truly cosmopolitan solution to climate change can come only 'from below' – that is, from movements beyond states – thereby contesting the fundamental logic of the climate change regime and capitalist relations of production that have contributed to causing climate change. While Felli acknowledges that cosmopolitanism tends to put human interests before those of the states, he believes that a simple opposition between states and individuals can be misleading. He argues that the opposition between human interests and those of the state in this context needs to be perceived in terms of the capitalist relations of production that suppose two separate spheres, namely economics and politics. Felli's chapter describes how this separation between economics and politics is played out at the international level where the existence of a world market is 'paralleled with the generalization of political sovereignties' in the form of a system of formally equal and autonomous states. International negotiations on climate change, and resulting agreements among states, are thus expressions of narrow national interests. This 'fetishist understanding of the international realm' is in turn an expression of capitalist relations of productions. Therefore, according to Felli, there is a need to go beyond the appearance of 'states interests' so as to expose the 'mediations that produce this fetishist understanding'. Felli looks at this problem, in the process answering the questions of why state interests appear to be opposed to human interests in the context of climate change, and what this tells us about the role of cosmopolitanism in addressing climate change.

Chapter 6, by Michael W. Howard, argues for 'qualified cosmopolitanism' to address climate change. Howard examines several proposals for principles that he believes should govern the sharing of the burdens of climate change. One idea he explores is that polluters should pay for the costs of climate change; indeed, he describes three versions of the polluter-pays principle. But, as he points out, this principle by itself is inadequate because it does not distinguish between poor polluters and rich polluters. He points out that burdens of climate change will fall heavily on the global poor, in at least two ways. First, the impacts of climate change will hit places inhabited by poor people who lack resources needed to adapt. Second, mitigating global warming requires reductions in carbon dioxide just as developing countries need to expand their energy to support development. According to Howard, justice obliges the world's wealthy countries to reduce greenhouse gas emissions at a rate that permits development in poor countries, and to assist these countries in meeting necessary greenhouse gas reductions of their own. He argues that the 'polluter pays' principle should be qualified by an 'ability to pay' principle. He makes a distinction between cosmopolitan (or global) justice and international justice, proposing something much more modest than unqualified cosmopolitan egalitarianism but more robust than the human rights minimum that non-cosmopolitans tend to favor. His position is cosmopolitan insofar as he takes persons, rather than states, to be the relevant *objects* of moral concern, but in focusing on principles that should govern the sharing of the burdens of climate change he also considers statist structures.

In Chapter 7, Robert Paehlke looks at cosmopolitanism and climate change in the context of the United States. Until recently the United States government has been unwilling to acknowledge much responsibility for climate change. Even today, with Barrack Obama in the White House, there is no assurance that the United States will act robustly to address the problem. Indeed, as Chapter 7 shows, many political leaders and citizens of the United States continue to oppose action on this issue. To understand this reluctance to act on US responsibilities, Paehlke examines the domestic politics of climate change. He points out that in the 1990s the US Senate expressly rejected the principle of common but differentiated responsibility among states for climate change because it placed a greater burden of immediate action on the developed countries – especially the United States. Given this history, it is not surprising that the Kyoto Protocol to the climate change convention has not been ratified by the United States, or that President Obama felt that he must proceed with great caution at the December 2009 Copenhagen conference of the parties. The question that Paehlke addresses is whether this extreme caution regarding climate change on the part of the United States can be reversed. In particular, he asks whether there is any prospect of a large number of Americans adopting a more cosmopolitan perspective, thereby recognizing the special obligations of those in wealthy nations, including the United States, regarding the global risks associated with climate change.

In Chapter 8 Philip S. Golub and Jean-Paul Maréchal determine that overcoming the 'planetary prisoner's dilemma' of climate change requires a cosmopolitan response. Golub and Maréchal's aim is to contribute to the theoretical debate by focusing on the issue of distributional justice among states and social classes. States have varying capabilities and different historical responsibilities in the context of climate change, a reality that derives from their historical role in the global political economy. As Golub and Maréchal point out, this issue featured prominently in the failure of the December 2009 Copenhagen conference to establish the foundations for a much stronger global climate change regime. At the same time, however, the rise of a large class of consumers in emerging economies requires a serious appraisal of the role of social classes in climate change. The purpose of new theorizing on this issue should be to help find ways to overcome the contradiction between global human needs and the present reality of the segmentation of the international system into discreet national units. The question for Golub and Maréchal is how trust-building mechanisms can be developed for effective international cooperation. Focusing on the roles of American and Chinese greenhouse gas emissions, their chapter examines the intellectual and normative challenge to imagining new forms of 'ordered pluralist cooperation' leading to convergence around common agendas that are in the overall human interest. Cosmopolitanism informs their suggestions for how to respond to this challenge.

In Chapter 9 Paul Harris proposes a way forward for the international climate change regime that acknowledges the responsibilities and duties of developed states while also explicitly acknowledging and acting upon the responsibilities of all affluent people, regardless of nationality. His aim is to explore the role of justice in the world's responses to climate change, and in particular to describe an alternative strategy for tackling climate change that is more principled and practical than the prevailing approach, and which may be much more politically acceptable to governments and citizens than are existing responses to the problem. This alternative strategy is premised on cosmopolitanism. A cosmopolitan ethic, and its practical implementation in the form of global justice, offers both governments and people a path to sustainability and successful mitigation of the adverse impacts of climate change. Harris's argument in favor of a more cosmopolitan approach to dealing with climate change is not meant to be an idealistic exercise or an act of imploring the world to accept that all people will quickly become good global citizens or that states can be abandoned in favor of cosmopolitan action alone. Rather, his argument is an attempt to show that the most practical and politically viable approach to climate change is in fact one that actualizes cosmopolitan ethics. Harris believes that, by placing persons, including their rights, needs and duties, at the centre of climate diplomacy and discourse, more just, effective and politically viable policies are more likely to be realized.

CONCLUSION

The world's responses to climate change may benefit greatly from taking a more cosmopolitan perspective because it is a global problem with global causes and consequences. As Held puts it, 'cosmopolitanism constitutes the political basis and political philosophy of living in a global age'.[77] The idea that states can continue to control what happens within their borders is no longer valid because 'some of the most fundamental forces and processes that determine the nature of life chances within and across political communities are now beyond the reach of individual nation-states'.[78] Climate change is one contemporary phenomenon that creates 'overlapping communities of fate' requiring new cosmopolitan institutions.[79] Where cosmopolitan justice is especially important is in locating obligation – to stop harming the environment on which others depend and to take steps to aid those who suffer from harm to the environment – on the shoulders not only of governments but also of capable individuals. As Attfield points out, 'the global nature of many environmental problems calls for a global, cosmopolitan ethic, and for its recognition on the part of agents who thereby accept the role of *global citizens* and membership of an embryonic global community'.[80] Cosmopolitan justice, and the associated obligations, might therefore at minimum supplement the traditional statist view of climate governance, although it should not dilute the common but differentiated responsibilities of states. This has significant implications for reshaping climate change policies and the climate regime more generally – in line with cosmopolitan and environmentally friendly objectives – without ignoring the reality of continuing state dominance of related institutions and policy responses.

NOTES

1. See, for example, Kevin Anderson and Alice Bows, 'Beyond "dangerous" climate change: emission scenarios for a new world', *Philosophical Transactions of the Royal Society A* 369 (13 January 2011): 20–44.
2. Here I am largely recounting the summary of cosmopolitanism in Paul G. Harris, *World Ethics and Climate Change: From International to Global Justice* (Edinburgh: Edinburgh University Press, 2010).
3. David Held, 'Principles of cosmopolitan order', in Gillian Brock and Harry Brighouse (eds) *The Political Philosophy of Cosmopolitanism* (Cambridge: Cambridge University Press, 2008), p. 10.
4. Thomas W. Pogge, *World Poverty and Human Rights*, 2nd ed. (Cambridge: Polity, 2008), p. 175.
5. Held, p. 12.
6. *Ibid.*, p. 15.
7. *Ibid.*
8. *Ibid.*, pp. 15–16.
9. Peter Singer, 'Famine, affluence and morality', in William Aiken and Hugh LaFollette (eds) *World Hunger and Morality* (Upper Saddle River, NJ: Prentice Hall, 1996); Simon Caney, 'Cosmopolitan justice, responsibility, and global climate change', *Leiden Journal of International Law* 18 (2005): 747–75.
10. Charles Jones, *Global Justice* (Oxford: Oxford University Press, 1999), p. 15.
11. *Ibid.*, p. 15.
12. Gillian Brock and Darrel Moellendorf, 'Introduction', *Journal of Ethics* 9 (2005): 2.
13. Andrew Linklater, 'Citizenship, humanity and cosmopolitan harm conventions', *International Political Science Review* 22 (3) (2001): 264.
14. Chris Brown, *International Relations Theory* (New York: Columbia University Press, 1992), p. 24.
15. Pogge, p. 175.
16. *Ibid.*
17. Charles Beitz, 'International liberalism and distributive justice: a survey of recent thought', *World Politics* 51 (2) (1999): 287.
18. Rainer Forst, 'Towards a critical theory of transnational justice', *Metaphilosophy* 32 (1/2) (2001): 164.
19. Simon Caney, *Justice Beyond Borders* (Oxford: Oxford University Press, 2005), p. 5.

20. Darrell Moellendorf, *Cosmopolitan Justice* (Boulder, CO: Westview Press, 2002), p. 172.
21. *Ibid.*
22. Gillian Brock and Harry Brighouse, 'Introduction', in Gillian Brock and Harry Brighouse (eds) *The Political Philosophy of Cosmopolitanism* (Cambridge: Cambridge University Press, 2005), p. 2.
23. *Ibid.*, p. 3.
24. Robin Attfield, *Environmental Ethics* (Cambridge: Polity Press, 2003), p. 162.
25. Brock and Brighouse, p. 3.
26. Thomas Pogge, 'Cosmopolitanism: A defense', *Critical Review of International Social and Political Philosophy* 5 (3) (October 2002): 86.
27. *Ibid.*, p. 87.
28. *Ibid.*
29. *Ibid.*, p. 89.
30. *Ibid.*, pp. 90–91.
31. *Ibid.*, p. 91.
32. Nigel Dower, *World Ethics*, 2nd ed. (Edinburgh: Edinburgh University Press, 2007), p. 28.
33. *Ibid.*, p. 11.
34. James Garvey, *The Ethics of Climate Change* (London: Continuum, 2008), p. 82.
35. Charles Beitz, *Political Theory and International Relations* (Princeton: Princeton University Press, 1979). While Rawls does not say so, several scholars, notably Brian Barry, Charles Beitz and Thomas Pogge, have argued that Rawls' premises lead to the conclusion that resources and wealth ought to be redistributed to the world's least well-off people. See John Rawls, *A Theory of Justice* (Cambridge, MA: Harvard University Press, 1971) and *The Law of Peoples* (Cambridge, MA: Harvard University Press, 1999).
36. Daniel Weinstock, 'Introduction', in Daniel Weinstock (ed.) *Global Justice, Global Institutions* (Calgary: University of Calgary Press, 2007), p. ix.
37. Dower, p. 92.
38. Wolfgang Sachs and Tilman Santarius, *Fair Future* (London: Zed, 2007), p. 125.
39. Onora O'Neill, *Bounds of Justice* (Cambridge: Cambridge University Press, 2000), p. 191.
40. *Ibid.*, p. 192.
41. *Ibid.*
42. *Ibid.*, pp. 195–6.
43. *Ibid.*, p. 199.
44. Andrew Dobson, 'Thick cosmopolitanism', *Political Studies* 54 (2006): 165–84.
45. *Ibid.*, p. 169.
46. *Ibid.*, p. 170.
47. Andrew Linklater, 'Cosmopolitanism', in Andrew Dobson and Robyn Eckersley (eds) *Political Theory and the Ecological Challenge* (Cambridge: Cambridge University Press, 2006).
48. Dobson, p. 182.
49. *Ibid.*
50. *Ibid.*, pp. 172–3.
51. *Ibid.*, p. 173.
52. *Ibid.*, p. 178.
53. O'Neill, p. 187.
54. Andrew Dobson, 'Globalization, cosmopolitanism and the environment', *International Relations* 19 (3) (2005): 270.
55. I describe cosmopolitan (or cosmopolitan-like) arguments, but not all of these scholars may necessarily call themselves cosmopolitans.
56. Lorraine Elliott, 'Cosmopolitan environmental harm conventions', *Global Society* 20 (3) (2006): 350.

57. *Ibid.*
58. *Ibid.*, p. 351.
59. *Ibid.*, p. 363.
60. Attfield advocates a consequentialist variant of cosmopolitanism based on needs. See Robin Attfield, *The Ethics of the Global Environment* (Edinburgh: Edinburgh University Press, 1999), p. 205.
61. Robin Attfield, 'Environmental values, nationalism, global citizenship and the common heritage of humanity', in Jouni Paavola and Ian Lowe (eds) *Environmental Values in a Globalising World* (London: Routledge, 2005), p. 41.
62. Derek Heater, *World Citizenship and Government* (London: Macmillan, 1996), p. 180.
63. Steve Vanderheiden, *Atmospheric Justice* (Oxford: Oxford University Press, 2008), p. 104.
64. Frank Biermann, 'Between the United States and the South', Global Governance Working Paper No. 17 (Amsterdam: Global Governance Project, 2005), p. 20.
65. Peter Baer, 'Equity, greenhouse gas emissions, and global common resources', in Stephen H. Schneider, Armin Rosencranz, and John O. Niles (eds) *Climate Change Policy* (Washington, DC: Island Press, 2002).
66. Jean-Charles Hourcade and Michael Grubb, 'Economic dimensions of the Kyoto Protocol', in Joyeeta Gupta and Michael Grubb (eds) *Climate Change and European Leadership* (London: Kluwer Academic Publishers, 2000), p. 199.
67. Simon Caney, 'Cosmopolitan justice, rights and global climate change', *Canadian Journal of Law and Jurisprudence* 19 (2) (2006): 255.
68. Caney, 'Cosmopolitan justice, responsibility, and global climate change', p. 768.
69. *Ibid.*
70. *Ibid.* Caney points out that knowledge of climate change has been sufficient since roughly 1990 to neutralize arguments of ignorance about its causes and severity.
71. *Ibid.*, p. 769.
72. Michael Mason, *The New Accountability* (London: Earthscan, 2005), p. x.
73. *Ibid.*, p. 12.
74. Dale Jamieson, *Morality's Progress* (Oxford: Clarendon Press, 2002).
75. *Ibid.*, p. 306.
76. *Ibid.*, pp. 306–7.
77. David Held, 'Principles of cosmopolitan order', in Brock and Brighouse, p. 27.
78. David Held, 'Regulating globalization?: The reinvention of politics', *International Sociology* 15 (2000): 399.
79. Vanderheiden, p. 89.
80. Attfield, *Environmental Ethics*, p. 182.

2. Climate justice as globalized responsibility: mitigation, adaptation and avoiding harm to others

Steve Vanderheiden[1]

INTRODUCTION

Who should pay the costs associated with anthropogenic climate change, how much should they pay, and why? This burden-distribution problem has become the central question of climate justice among scholars and activists and remains the primary obstacle to the development of an effective climate regime in practice.[2] These costs are expected to be significant and varied, but can generally be categorized in terms of *mitigation*, or those costs associated with reducing further human contributions toward the increasing atmospheric concentrations of heat-trapping greenhouse gases (GHGs) that cause climate change, and *adaptation*, or those costs associated with attempting to insulate humans from climate-related harm from existing anthropogenic environmental damage.[3] Since mitigation actions undertaken by developed countries under the auspices of the Kyoto protocol are self-financed and mitigation targets accepted by developing countries are widely viewed as contingent upon financing from developed countries, imperatives to reduce GHGs are fundamentally matters of allocating mitigation costs. Properly speaking, adaptation intervenes in the causal chain between climate change and human harm, allowing the former but preventing the latter, but when this is not possible a third category of *compensation* costs must be assigned in order to remedy failed mitigation and adaptation efforts. Because the formulae for assessing liability for adaptation and compensation are identical,[4] and since climate justice requires adaptation efforts that render compensation unnecessary,[5] for the purposes of this chapter the category of adaptation shall be understood to include both *ex ante* prevention of harm as well as *ex post* compensation for it. As expected, the Copenhagen Accord that emerged from the 15th conference of the parties (COP-15) of the 1992 United Nations Framework Convention on Climate Change (UNFCCC) in December 2009 failed to satisfactorily address this core burden allocation issue,[6] making its resolution the primary problem at COP-16 in Cancun at the end of 2010.

Sufficient mitigation actions must stabilize atmospheric concentrations of carbon dioxide at levels that 'avoid dangerous anthropogenic interference' with the planet's climate system, as declared by UNFCCC and quantified at COP-15 as resulting in no more than a 2 degree Celsius global temperature increase. To achieve this goal, GHG emission reductions of approximately 80 percent from 2000 levels will be needed by 2050, and such reductions will require significant infrastructure investments and/or foregone consumption, even if these actions also yield long-term net benefits.[7] Likewise, the UNFCCC estimates adaptation costs at between US$40 and US$170 billion per year, which some critics suggest is significantly underestimated.[8] Whatever the total costs of sufficient mitigation and adaptation efforts, these costs must be fully assigned and undertaken if climate injustice is to be avoided, for to fail in mitigation is to allow catastrophic environmental damage, and to fail in adaptation is to wrongfully allow avoidable human suffering to occur. The human community must ask and answer this question, for as Simon Caney writes, 'we cannot accept a situation in which there are such widespread and enormously harmful effects on the vulnerable of this world'.[9] If we do not act in accordance with our answers, the way those costs will be allocated by the global calamity of unmitigated climate change will be inexcusably unjust, and will very likely be worse than even the most misguided remedial efforts.[10]

To be effective, a global climate policy must be accepted by all national parties, and to be acceptable to all – essential in the absence of a coercive world regulatory state that can ensure compliance through force alone – it must be fair to each. For such reasons, philosophical inquiry into justifiable burden allocation formulae is an eminently practical exercise, as the most defensible allocation formula is no more important than its reasoned justification. Philosophers and political theorists most commonly turn to principles of distributive justice in an effort to give a principled account of justly allocated burdens or costs, which is understandable in view of the fundamentally distributive problem that burden allocation typically entails. Following such an approach, scholars ask how to equitably allocate shares of atmospheric absorptive capacity, which allows for greenhouse emissions without deleterious effects upon climate, typically endorsing some variation upon the equal per capita emission rights (EER) view, to be explored below. Some, by contrast, invoke corrective justice on behalf of such problems, arguing that responsibility rather than equity ought to be the guiding principle for assigning climate-related costs, typically assigning proportionally larger burdens to bigger historical greenhouse polluters, as Michael Howard does in his chapter. I have previously endorsed a hybrid view, arguing that distributive justice theory provides fitting principles for assigning the costs associated with mitigation, but that an account based in corrective justice supplies those principles applicable to adaptation. I'm still convinced of the need for this distinction, given the separate normative issues raised by these two sets of problems.[11] Because these two imperatives of climate justice rest on different conceptions of justice and involve fundamentally different forms of normative analysis, my concern here is whether or not they can be reconciled under one overarching view of climate justice, and indeed whether justice theory itself can reconcile the demands of distributive and corrective justice.

Largely because of the refusal by developed countries to accept any formula for assigning adaptation burdens other than voluntary contributions by presumably charitable parties, none of the international agreements made under the auspices of the UNFCCC has provided grounds for associating mitigation and adaptation responsibilities. Since liability for adaptation has not been tied to past, present or projected future national emissions, deficiencies in mitigation efforts have no effect on ongoing adaptation responsibilities under current international policy frameworks. This is objectionable from the perspective of corrective justice, however, since failure to undertake one's assigned mitigation burdens results in greater fault-based liability for climate related harm, which should result in greater adaptation burdens. This burden-allocation policy problem reflects a deep theoretical incommensurability between the requirements of distributive and corrective justice, on which climate change-related mitigation and adaptation imperatives are based. Both are seen as aspects of justice writ large, but their fundamentally different structures complicate the parsimonious combination of distributive and corrective justice within a single conception of justice, as mitigation and adaptation imperatives based upon them have likewise proven difficult to combine within a single climate justice metric. But they must be combined in some way, insofar as a responsibility-based account of adaptation liability depends upon recent past mitigation efforts, as I have argued they must.[12] For reasons which will be sketched below, greater recent effort at mitigation ought to reduce future adaptation burdens as a matter of justice, but this draws upon a comprehensive notion of justice that is able to successfully combine its distributive and corrective aspects.

Exacerbating this theoretical incommensurability is the disparate impact of mitigation and adaptation activities in practice. Because they involve different kinds of activities and stand to benefit different groups of persons, with mitigation yielding primarily global benefits from diminished climate disruption and adaptation producing local benefits from specific projects, any commensurability in terms of costs would not be reflected in terms of benefits. Should nations be allowed to rectify insufficient mitigation efforts with increased adaptation activities, their total costs might be held constant but the beneficiaries of their combined activities would not be the same. Especially if allowed to count domestic adaptation activities toward combined mitigation and adaptation burdens, nations could continue to cause global harm through their inadequate mitigation activities while shielding only their domestic populations from climate impacts, clearly transgressing the demands of climate justice. Insofar as justice is concerned with the allocation of benefits as well as the assignment of burdens or costs, the use of a single metric for calculating and discharging climate-related remedial obligations ignores this problem of benefit distribution. It would seem that climate justice requires fully adequate action in mitigation *and* in adaptation, but this again raises the non-ideal theory question of how national inadequacy in performance of one climate justice imperative affects the ongoing assignment of burdens in the other, which is of tremendous policy relevance in ensuring that the normative objectives of global climate policy are achieved.

I shall argue that the normative concept of responsibility offers the value basis for linking mitigation and adaptation efforts under a single overarching conception of justice, transcending the distributive and corrective conceptions noted above and providing a coherent account of climate justice capable of resolving the difficulties noted above. This view of climate justice as being linked by an account of responsibility can be stated in brief: justice demands that persons and peoples voluntarily take or be made to bear responsibility for all and only the climate change that they culpably cause, or be held responsible for it by others. If this can be done for all persons and peoples that affect or are affected by climate change, then climate justice can be usefully understood as an effort at ensuring globalized responsibility. Being responsible, in this sense, requires that persons and peoples avoid harming others through the environmental externality of anthropogenic climate change, whether by paying the relevant mitigation costs needed to avoid causing climate change or by paying the adaptation costs need to avoid this resulting in human harm. Insofar as persons and peoples fail to do their share in mitigating this global environmental problem and/or controlling its effects, they can be held responsible by others through assessments of liability to pay such costs, or through compensation orders. To the unpacking of this claim and its linking conception of responsibility we must now turn.

MITIGATION VERSUS ADAPTATION

Mitigation efforts involve assigned reductions to human emissions of greenhouse gases, enhancement of carbon sequestration capacities, or both. The essential policy tool of such efforts is the GHG emission cap – or, viewed positively in terms of allowances rather than negatively in terms of constraints, the emission right – which captures both approaches. To minimize further anthropogenic contributions toward climate change, nations or persons must comply with caps that measure net emissions, counting the effects of initial GHG emissions into the atmosphere as well as their sequestration in sinks like forests or underground storage facilities. Within the net allowable annual GHG emissions associated with this primary objective, it is (unlike burdens associated with adaptation efforts) fundamentally concerned with justly distributing the common resource of atmospheric absorptive capacity needed to accommodate ongoing human greenhouse emissions without deleteriously affecting global climate. Its fundamental normative question is thus: how much of this finite, common resource is each of us entitled to claim through our GHG-emitting activities? Viewing such entitlements as rights, we can pose the same question in another way: At what point do we exhaust our emission rights and begin to wrongfully produce excessive emissions, for which we may be liable? Climate justice imperatives demand that emission caps eventually be set at such sustainable thresholds and may temporarily require caps to be set below such thresholds in order to decrease atmospheric concentrations of GHGs.[13]

On the basis of cosmopolitan justice principles such as those developed by Paul Harris in his chapter, most scholarly commentators defend some version of the EER thesis, arguing that all persons are entitled to equal shares of atmospheric absorptive capacity, such that national emissions caps should be calculated on an equal per capita basis. Sometimes, a modified version of EER is defended with minor deviation from this equal per capita standard, taking into account geographic differences that influence national energy consumption patterns or controlling for disparity in benefits derived from domestic renewable energy resources or carbon sinks. Others defend a version of EER over a long period of compliance, such that higher past national per capita emissions must be offset by lower caps in the future.[14] My own view assigns national emissions caps on a modified EER basis but in terms of equitably allocated luxury emissions, defined in contrast to the survival emissions that are required to meet basic human needs, to which I claim that persons have rights.[15] Regardless of the version of the EER thesis, all such approaches treat mitigation as fundamentally a distributive justice problem that requires egalitarian distributive principles in order to find a solution. The question of how much of this shared resource each nation or person is entitled to claim if climate change is to be sufficiently mitigated is categorically different from the question of what to do if we together fail to avoid that problem. The latter question is one for corrective justice, and shall be considered below.

Greenhouse emission caps entail burdens for those that currently emit greenhouse gases at levels that exceed them, and benefits or opportunities for those whose emissions are currently at rates below those assigned under a cap (or cap-and-trade) system. Hence, any version of EER will by necessity entail larger burdens for bigger current greenhouse polluters, since they will be required to engage in more significant emissions reductions in order to comply with those caps. But the aim of mitigation is to minimize further contributions to the problem, not to assign fault and assess liability for past actions, and its primary task is to assign fair shares of a common and finite but currently overappropriated resource. The 'historical responsibility' approaches to EER, that look to high past national emission rates in order to justify lower future emission caps, do so from distributive and not from corrective justice, and are typically silent on adaptation questions. For all versions of EER, past actions are only relevant to currently-assigned mitigation burdens insofar as they define the gap between existing emission rates and the fair shares of atmospheric resources around which future caps are set, which makes those actions relevant to the costs of compliance with those caps but not to the targets themselves. Since they merely extend the compliance period for equitably-assigned per capita caps, even the ostensibly punitive historical responsibility approaches deny that past actions are relevant to the assignment of current and future emission targets, though they are relevant to determining compliance with those targets.

Distributive justice is forward-looking and based in equity, but is not remedial in regards to past inequity, and so requires a corrective component in order to rectify past injustices. In the context of climate change, this remedial aim is not satisfied by merely extending the temporal scope of distributive justice obligations. Distributive justice offers principles that define a just allocation of GHG emission rights, setting a target toward which near-future mitigation actions must aim, but those principles yield no directives for what to do when their requirements are violated. Climate change mitigation is likewise distributive but not fundamentally remedial – it looks forward rather than backward, so to speak – and so requires a remedial component to rectify past and ongoing mitigation failures. Backward-looking adaptation measures, based in principles of corrective justice and aimed at prevention of or compensation for harm rather than the rectification of past distributive injustice, provide this necessary remedial justice complement to forward-looking mitigation measures, and must be kept conceptually distinct in theorizing climate justice and constructing policy responses from such theory. But this again raises the question of how two distinct climate justice imperatives based in distinct conceptions of justice can be combined into a single burden allocation formula that is capable of simultaneously ensuring that sufficient mitigation and adaptation activities are prescribed, reconciling the distributive and corrective principles on which the assignment of such costs are based.

DISTRIBUTIVE JUSTICE AND REMEDIAL RESPONSIBILITY

Within the ideal theory literature in which distributive justice theories have primarily been developed, which assume compliance with the terms of justice principles, the remedial aspect of justice is often slighted in favor of development and defense of principles defining just distribution or discussions identifying the goods that are subject to such principles. Justice is defined in terms of distribution rather than *re*distribution, despite the obvious problem that an initially just distribution of primary goods can become unjust in short order in the absence of state intervention. A complete account of justice entails principles for just distribution and a process whereby departures from those principles can be appropriately addressed and corrected.

In order to flesh out the conceptual links between the normative bases of mitigation and adaptation imperatives, I draw upon a view of responsibility drawn from luck egalitarian theory for its instructive emphasis on remedies that restore distributive justice as various events disturb this equilibrium over time. The core premise concerns the link between voluntary control, responsibility and entitlement: it presumes that control can be the source of entitlement and moral responsibility but that factors outside of an agent's control (defined as *luck*) cannot. Upon noting the causal role of luck in determining part of a person's opportunity structure, John Rawls describes the role of distributive justice as providing a social remedy to unequal natural allocations that are 'arbitrary from a moral point of view',[16] but he never explores the flip side of responsibility, where control over one's circumstances might lead to justified inequality. Noting this asymmetrical view on luck, responsibility and entitlement, others in the egalitarian tradition have explicitly maintained that persons deserve all and only those good or bad outcomes that result from their voluntary acts and choices, but not those which result from luck alone. G.A. Cohen, for example, interprets the Rawlsian principles in such a way,[17] and Ronald Dworkin distinguishes between deserved and undeserved sources of inequality based in the causal role of luck in those outcomes, where persons are held to be responsible for those consequences resulting from their deliberate choices but not their native endowments.[18] However, luck egalitarian justice theories contain no corrective justice component, relying instead upon periodic redistribution of resources to maintain distributive equity over time. Since such approaches find no direct link between distributive injustice and harm to others – indeed, the injustice that they identify involves no interpersonal harm or injury, but instead involves violation of entitlement – the account of justice that they develop requires only that equity be restored by neutralizing the effects of luck on holdings, not that responsibility for harm be established or that compensation be provided by culpable parties. Such theories, however, suggest the overarching account of responsibility that can provide a conceptual bridge between the demands of distributive and corrective justice, even if they never fully develop it. To develop it here, and thereby to conceptually link the normative bases of mitigation and adaptation imperatives, corrective justice-based theories of responsibility will be canvassed for their use of this link between voluntary action, harm and the remedial measures of corrective justice.

LUCK, DIFFERENTIATED RESPONSIBILITIES AND JUSTICE

So what does luck have to do with responsibility and justice? How do these concepts together inform the question about allocating climate-related costs? To begin, we might notice that the language of the UNFCCC emphasizes responsibility as a core element of a just response to climate change. In its oft-quoted Article 3.1, that treaty declares that ‘the Parties should protect the climate system for the benefit of present and future generations of humankind, on the basis of equity and in accordance with their common but differentiated responsibilities and respective capabilities’. As the earth’s climate system itself is common to all, so also is the responsibility to protect it. But that doesn’t mean that the (forward-looking) responsibility to mitigate climate change is to be equally shared. The phrase ‘differentiated responsibilities’ is here a reference to the judgment that states with the largest historical and current emissions are more causally responsible for the problem and ought therefore to bear primary remedial or liability responsibility for its mitigation. In this context, at least, it appears not to be a reference to liability for adaptation, as the focus is on the assignment of burdens needed to ‘protect the climate system’ and not what is necessary to insulate humans from harm that results from climate change, but the 'common but differentiated responsibilities' idea could equally well apply to adaptation efforts.

Worth noting is the 'and respective capabilities' phrase that follows, since this implies a quite different burden-sharing formula. Both 'equity' and 'differentiated responsibilities' imply that remedial responsibility be assigned in proportion to causal responsibility or in accordance with standards of strict liability, as in the polluter-pays principle: that big historical and/or current greenhouse polluters ought to have proportionally larger burdens to bear than smaller polluters. But 'respective capabilities' implies an ability-to-pay criterion, like Caney and Howard include in their hybrid accounts. Neither criterion implies that developing countries should be fully excluded from participation in a system of mandatory GHG caps, as some claim, even if these normative criteria are pragmatically consistent with excluding them from the first round of caps and timetables. Absent some distinction between survival emissions, to which persons have rights and which are therefore seen as not responsible for climate change, and luxury emissions to which they have no such rights, combined with an empirical claim that developing countries have not yet and will not in the future emit beyond this survival threshold, there is no justified ground for a permanent exemption for the global South from mandatory caps. As I have argued elsewhere,[19] they must be included under a system of mandatory caps, else the growth of emissions in the South risks undermining any decarbonization efforts in the North.

Given the above distinctions between mitigation and adaptation efforts and the normative principles that are most appropriate to each, the question arises: why did the UNFCCC invoke the language of 'differentiated responsibilities' in the context of climate change mitigation rather than its adaptation? In what follows, I mean to highlight the role of responsibility as an overarching conception that links together the duties of mitigation and adaptation under a just climate regime, and which simultaneously provide the content for adaptation efforts. The term responsibility carries with it various distinct if sometimes overlapping meanings, and before we can specify what it has to do with justice we must sort out some of these meanings. I suspect that delegates at the time wished to avoid reference to a remedial component that might require industrialized countries to fund adaptation and compensation efforts, explaining this context, but now the responsibility language can be seen as the normative framework that links the 'protect the climate system' mitigation language with the 'benefit of current and future generations of humankind' language that also captures adaptation imperatives.

This relationship between justice and responsibility can be illuminated by considering the various related ways in which agents (whether persons or groups) can *be* responsible. To *take* responsibility is to voluntarily take ownership of or be answerable for the positive and negative consequences of one's actions. Regarding negative consequences, one who takes responsibility for their actions takes due care to ascertain and publicly acknowledge any problems toward which she contributes, and is committed to taking reasonable measures to minimize one's further contributions to them and (in some cases, at least) to rectify any harm that has already been done. One can also take responsibility for the beneficial consequences toward which one contributes, though such credit-claiming is not a core obligation of ethics or a component of justice. For the sake of symmetry, though, responsibility for outcomes involves playing a role in the causal narrative that generates those outcomes in a manner that is similar regardless of whether they are good or bad for others. Agents can be *held* responsible for those same good or bad consequences when they fail to voluntarily own up to those consequences themselves but are subsequently ordered to so by some third party. As above, justice is primarily interested in holding parties responsible for bad consequences, though there may be some social interest in holding them responsible for good ones, too.

To *be* responsible in this sense – whether voluntarily or by imposition, and whether proactively through harm-avoidance or retrospectively through remedial justice – is to comply with the obligations that normatively constrain action, and here the relationship between responsibility-based justice and ethics can more explicitly be drawn. Agents have an *ethical* obligation to refrain from causing foreseeable and avoidable harm to others, and this has priority over the remedial duty of justice that arises when they fail to comply with its imperatives. When agents fail to observe this primary negative ethical duty, they incur a remedial duty to rectify the wrong that they have done to their victim. At this point of offense, justice demands that they either voluntarily take responsibility or involuntarily be held responsible for providing an adequate remedy. Such remedies can include compensation or other forms of reparations, but may also include apologies and other public gestures of contrition or (especially when victims cannot themselves be fully restored to justice) efforts to aid victims of similar offenses or prevent future like offenses from occurring. For example, someone who causes a fatality while drunk driving may not be able to rectify this injustice with its victim but might instead work to assist other victims of impaired drivers or seek to prevent intoxicated drivers from getting behind the wheel in the future. While voluntarily taking responsibility is considered to be more admirable than reluctantly being compelled to perform some remedial action, the external content of these two forms of rectification is otherwise identical. Obligations of remedial justice arise, that is, when those ethics of nonmaleficence are breached.

The two most common forms of remedial justice are retributive and restorative justice. As a theory of punishment, retributive justice supposes that an offense by a perpetrator against a victim creates some kind of cosmic imbalance between the two that cannot be removed without visiting a similar injury upon the perpetrator, typically by the state but potentially also by other authoritative agents. The classic 'eye for an eye' account of the nature and justification punishment takes this view, where punishments are supposed to be proportional to the original offense in order to restore this balance between perpetrator and victim. Note that retributive justice does not attach any particular value to assisting the victim, however, and regards society rather than the victim as having the claim to extract the appropriate remedy. The practice of pitting criminal defendants against the state as opposing parties in legal proceedings is a legacy of this retributive theory of punishment, and conceives of punishment as 'paying one's debt to society' rather than retiring any debt to the victim as the essence and justification of punishment. Those responsible for bad outcomes are held to account for their role in generating such outcomes through retributive processes, so this variety of remedial justice is appropriate only for bad outcomes for which relevant agents cannot or do not take voluntarily responsibility. Finally, retributive justice is often seen as appropriate only for violation of publicly recognized and thus legally enforceable rules or norms, rather than for rectification of bad outcomes as such, as retribution is seen as vindicating and reinforcing those rules or norms. Indeed, legal positivists deny that rules or norms have any valid existence in the absence of such occasional enforcement.

Restorative justice likewise conceives of injustice as creating a kind of imbalance between perpetrator and victim, but views the essence of the offense as against the victim or community rather than the state. Responsibility in the restorative sense thus involves a process of acknowledging this damage and seeking to repair it, and can for this reason be voluntarily taken as well as coercively imposed. In contrast to retributive justice, then, restorative justice requires that someone (the victim) be helped rather than that someone else (the perpetrator) be harmed as the core remedy, and aims to restore relationships of solidarity between the two rather than enacting a kind of socially-sanctioned revenge as antidote to socially-dispersed offense. Because its focus in on restoration of a desirable state of affairs that had been disrupted by some offense, restorative justice looks forward in a way that retributive justice does not, and doesn't necessarily depend upon judgments of fault, since it is not punitive. One can bring about an imbalance in need of restorative remedy without the *mens rea* required of criminal culpability, as in accidentally caused harm, since the need for a remedy is viewed from the perspective of victims. Perpetrators need not unjustly gain at the expense of their victims, but rather become indebted to them as the result of harm that the latter are unjustly made to suffer. Restorative justice thus is not a theory of punishment, although restorative processes are seen as a viable alternative to retributive punishment within criminal justice and have obvious utility for international law conflicts in which no adequate executive authority exists to enforce normative claims to rectification of prior offenses.

The conception of responsibility that operates within climate justice is a hybrid of these two forms of remedial justice. Judgments of fault are both possible and necessary where parties act in bad faith or otherwise fail to comply with their commitments, in order to give the climate convention positive force and overcome the collective action problems associated with burden allocations that are perceived as merely voluntary and thus not enforceable. Some centralized authority must be able to determine compliance and punish noncompliance with the terms of the convention, especially if GHG emission markets are to function effectively, and retributive justice theory provides justification for these elements of a responsibility-based justice process. But it is not sufficient to assign liability to those parties that exceed their fair shares of GHG emissions if the proceeds from this assessment are not redirected to the victims of climate change, whether through adaptation or compensation. Restorative justice theory contributes this victim-based and forward-looking perspective, and underscores the need for solidarity and cooperation in addressing the 'common but differentiated responsibilities' of all in protecting the earth's climate system that the Framework Convention rightly notes. Vindication of a nascent global climate regime and its fledgling regulatory power is a necessary means to climate justice, but its proper end is the maintenance of justice and its restoration where necessary and possible, and cooperation is as essential as coercion.

FAULT, RESPONSIBILITY AND LIABILITY

The overarching view of responsibility that encompasses the distributive and corrective justice elements of climate change mitigation and adaptation imperatives depends upon three related meanings of the term. The first concerns causation, wherein a person or group is said to be responsible for some outcome if they are a necessary cause of that outcome, whether through their acts or omissions. Whereas this sort of responsibility appears to be purely empirical and relatively straightforward, in the case of complex chains of causation like those in global climate change it can raise problematic questions about agency and causality. The second concerns a moral judgment about fault and assessment of liability, and typically depends upon causal responsibility as a necessary but insufficient condition. Third is remedial responsibility, wherein responsible agents are required to do something in response to past outcomes for which they are responsible. Specifically, the form of remedial responsibility relevant to climate change cost allocation is that which justifies assessments of liability to pay damages, and does not concern apologies, agent regret or contrition, criminal culpability or liability to punishment or the fitness of moral praise and blame. This limited purview is justified by the task at hand: we need to know what responsibility theory can say about how to allocate the cost of climate change. Whether or not people should feel guilty about contributing toward climate change, or apologize for it, is beyond the scope of this chapter. If the core imperative of climate justice is ensuring that those not responsible for causing climate change be insulated from having to bear its costs – as I claim that it is – then such ancillary questions about other forms of responsibility are beside the point.

These three models are causal, liability and remedial responsibility, respectively, and are distinct but interrelated in ways that the remainder of this section will endeavor to explicate. Joel Feinberg describes the relationship between the first two in noting the conditions in this standard legal model of liability based in *contributory fault*:

> First, it must be true that the responsible individual did the harmful thing in question, or at least that his action or omission made a substantial causal contribution to it. Second, the causally contributory conduct must have been in some way *faulty*. Finally, if the harmful conduct was truly 'his fault,' the requisite causal connection must have been directly between the faulty aspect of his conduct and the outcome. It is not sufficient to have caused harm *and* to have been at fault if the fault was irrelevant to the causing.[20]

According to this model, agents are responsible for the bad consequences that they cause through their faulty action, and so are liable for providing the appropriate remedy. But when applied to climate change, this liability model of responsibility breaks down. It is difficult if not impossible to establish direct causality between any individual GHG-emitting act and harm, as climate-related harm is the cumulative result of a great many separate emissions rather than any discrete action committed by any distinct person. Thus, it is difficult to identify any non-circular account of how individual acts can be faulty, particularly since many polluting acts are widely tolerated and even encouraged by existing social norms.

In its standard sense, at least, it does not seem that persons can be held responsible for climate-related harm through fault-based liability.

These problems with the liability model of responsibility have been noted elsewhere. For example, Robin Attfield describes such problems as creating a kind of 'mediated responsibility' for climate-related harm, which he argues ought nonetheless to be regarded as sufficient for establishing fault and, by extension, ascribing liability. He finds the ethical theory from which responsibility is defined to be deficient in such complex cases, urging that 'the scope and content of ethics, then, must be reconceived so as to correspond to the full range of the foreseeable impacts of human activities'.[21] Others are less sanguine about the applicability of the liability model to complex cases, however. Writing about the responsibility of consumers for sweatshop labor conditions but in a description that could also be applied to individual responsibility for climate change, Iris Young identifies the fragmented causality and diffused agency that undermine assignments of responsibility in this contributory fault model. This model, she notes:

> relies on a fairly direct interaction between the wrongdoer and the wronged party. Where structural social processes constrain and enable many actors in complex relations, however, those with the greatest power in the system, or those who derive benefits from its operations, may well be removed from any interaction with those who are most harmed in it.[22]

Young argues that it is 'usually inappropriate to *blame* those agents who are connected to but removed from the harm', but blaming is not the same as finding fault. Sweatshops, like climate change, are examples of what Young terms *structural injustice*, which often yields 'collective results that no one intends, results that may even be counter to the best intentions of the actors' and in which 'in most cases it is not possible to trace which specific actions of which specific agents cause which specific parts of the structural processes or their outcomes'.[23] If structures play some causal role in such harmful phenomena, the role played by individual agents may be thereby diminished, along with the grounds for individual culpability.

But Young does not mean to exonerate individuals from responsibility for harm that results from structural injustice. Rather, she argues, by participating in such structures – for example by purchasing sweatshop-produced garments or emitting excessive greenhouse gases – agents are 'producing and reproducing the structures' and hence 'are implicated in that responsibility'.[24] Diffusion of moral responsibility doesn't cause it to disappear altogether, although it does blur the distinction between culpable and innocent agents. So while the direct links between faulty conduct and harmful outcomes required by the liability model may be impossible to establish, this does not diminish individual culpability, even if it does complicate it considerably. These complications lead Young to eschew liability responsibility in favor of a 'social connection' model of social and political responsibility, which 'allows us to take responsibility *together* for sweatshop conditions, without blaming anyone in particular for the structures that encourage their proliferation'.[25] This form of collective fault-based responsibility 'adds to rather than replaces this first layer of responsibility'[26] and so does not disallow findings of individual fault where lines of causation between individual acts and associated harm can be established. Her motive for assigning fault collectively is largely pragmatic, and applies equally well to climate change: if persons are held to be collectively rather than individually responsible for structural injustice, their appropriate response would thus also be collective rather than individual, carried out through political rather than private action. Since no one 'can act alone to improve working conditions', individual assessments of liability cannot provide such an effective remedy.

Working within the corrective justice paradigm of tort law, Matthew Adler doubts that individual liability for climate-related harm can be fit within existing legal norms. For one thing, he notes, US tort law focuses on 'personal injury or property harms' rather than 'losses to well-being per se' or 'pure economic loss', but the expected harms associated with climate change include such collective damage as 'sea level rise, harm to natural systems such as coral reefs or glaciers, and drought or loss of water supplies' which 'are not themselves losses to individuals' paradigmatically protected interests'.[27] If tort damages were to be based on personal rather than collective harm, he argues, the direct causation requirement of fault-based liability would pose the obstacles noted above, since 'the causal links between a particular set of GHG emissions and those protected interests will generally be more attenuated that the links between those emissions and environmental damage'.[28] Moreover, torts offer a remedy to existing but not possible future harm, and 'tort law generally does not compensate for pure risk imposition'. For such reasons, Adler suggests a novel form of collective fault-based liability for collective rather than individual climate-related damage, through a new tort mechanism based on 'compensation by governments to other governments for past (not expected) environmental damage'.[29] Like Young, Adler finds group fault and liability to be preferable to individual judgments concerning responsibility, based on its conformity to existing legal norms rather than Young's pragmatic political reasons. Both, however, agree that diffused responsibility does not diminish fault so much as weigh in favor of a moral accounting system whereby it is ascribed collectively rather than individually.

Finally, remedial responsibility concerns itself with identifying appropriate responses to assessments of fault-based liability, as Howard does with the 'qualified cosmopolitanism' that he develops in his chapter. To be consistent, it must make a distinction between harm that results from faulty as opposed to faultless acts, ordering payment of damages only in cases involving the former. Here, Barry summarizes what he takes to be the core tenet of egalitarian distributive justice in a formulation that links distributive with corrective justice:

> A legitimate origin of different outcomes for different people is that they have made different voluntary choices . . . The obverse of this principle is that bad outcomes for which somebody is not responsible provide a *prima facie* case for compensation.[30]

Notice that Barry's principle refers to individual persons and not to groups, so it is grounded in a theory of individual but not collective responsibility. As previously noted, this formulation subsumes remedial responsibility within distributive justice rather than maintaining an independent account of corrective justice that is capable of addressing the sort of fault-based liability needed for assigning the costs of climate change adaptation, but its reference to grounds for compensation suggests how fault-based liability may be applied. For Barry, compensating the effects of bad luck is a social task, as when those born into poverty or stricken by disease are due redistributive transfers, and assigning individual liability for such compensation is unnecessary when bad luck for victims does not result from the faulty acts of others. Distributive injustice, by this account, is a function of unequal outcomes in the absence of responsibility for them. By invoking a luck egalitarian basis for principles of distributive justice – claiming that persons are entitled to all and only those benefits and burdens that result from their voluntary choices rather than those resulting from acts or events for which they are not responsible – Barry suggests how a single account of responsibility can generate distributive and corrective justice principles, because what looks like bad luck from the perspective of its victim can sometimes be attributed to culpable choice from the perspective of its perpetrator.

Caney notes that mitigation and adaptation imperatives involve 'two distinct kinds of burden' to allocate,[31] but nonetheless proposes allocating both kinds according to the same formula. His hybrid mitigation burden allocation formula combines fault-based liability (exempting pre-1990 emissions as the product of 'excusable ignorance') with an 'ability to pay principle' that requires 'the most advantaged' individuals to engage in either mitigation or adaptation activities up to the point that none is exposed to climate-related harm. But Caney's hybrid approach conflates equity-based concerns for distributive justice with responsibility-based ones for corrective justice, conflating two distinct standards of liability without acknowledging their divergence in justification or effect. Although both would assign mitigation and adaptation primarily to the affluent, and in rough proportion to their affluence, one does this on the basis of fault for causing climate change, while the other depends upon comparative burdens of assistance. Rather than applying distinct justice principles to these 'two distinct kinds of burden', he applies both kinds of principles to the sum of both kinds of burdens, in effect faulting affluence itself rather than unsustainable expressions of it by severing fault from liability, and failing to specify how deficiencies in mitigation affect the assignment of adaptation burdens.

After noting that the most advantaged 'may not have caused the problem' by virtue of their affluence alone, he suggests that they may have other reasons for being responsible for solving the problem. Citing Singer's argument for a duty to relieve famine[32] on behalf of this standard, Caney suggests that capacity to assist might justify such liability. But Singer's drowning child case involves a potential harm for which none is actively responsible, whereas climate change involves active harm and so allows for liability to be attached to fault as well as capacity to assist. In distributing remedial burdens among multiple parties, good reasons based in luck egalitarian principles and the corrective justice framework canvassed above support basing it on fault rather than mere capacity, though of course the child should be rescued by the capable if the faulty are unable to do so. Moreover, Singer's example involves a single would-be rescuer, who would arguably cause the child's death by an omission if she failed to act, whereas the climate case involves many would-be rescuers.

A better analogy to climate change would be Feinberg's case of a man drowning within audible distance of a thousand accomplished swimmers, who all leave him to drown. Illustrating collective rather than individual fault-based liability, Feinberg argues that 'each has a duty to attempt rescue so long as no more than a few others have already begun their efforts',[33] but the example is meant to illustrate how liability can be distributed among group members according to their relative fault under the form of collective responsibility that he terms *contributory group-fault: collective and distributive*. Those closer and thus in a better position to help bear greater liability and those further away may be less liable, Feinberg suggests, but all are responsible and hence liable for paying damages in proportion to their liability. Fault-based liability is preferable to capacity-based liability where degrees of fault are readily ascertainable, as they are with climate-related harm, so the former offers a superior standard for assigning adaptation costs in a way that is distinct from that relevant to mitigation. All that remains, then, is a conceptual bridge between the corrective justice account of fault-based liability that is used to assign adaptation burdens, and the distributive justice account of responsibility to do one's share to mitigate climate change or refrain from using more than one's share of atmospheric absorptive capacity.

CONCLUSIONS: GLOBALIZING RESPONSIBILITY

Returning to the burden allocation question that began this chapter, we must ask how this responsibility-based conception of justice would allocate the costs associated with climate change, then consider whether it gives a superior account to its alternatives. If we assume, as the account of justice that I have sketched above does, that persons should not be subjected to climate-related harm for which they are not themselves responsible, then we must begin with the strong imperative to avoid causing climate change. This is an imperative of justice, based on the negative responsibility obligation to avoid causing harm. In this sense, it is of the most robust variety of justice-related obligations, potentially giving rise to what Andrew Dobson calls 'thick cosmopolitanism' as described as follows:

> Causal responsibility produces a thicker connection between people than appeals to membership of common humanity, and it also takes us more obviously out of the territory of beneficence and into the realm of justice. If I cause someone harm I am required as a matter of justice to rectify that harm. If, on the other hand, I bear no responsibility for the harm, justice requires nothing of me – and although beneficence might be desirable I cannot be held to account (except in the court of conscience or God) for not exercising it.[34]

All persons have widely-recognized and basic rights not to be harmed, so this primary imperative is grounded in an uncontroversial injunction that is justified by any plausible normative theory. Meeting this objective requires some allocation of mitigation burdens, in the form of GHG emission caps. These caps should be allocated first among the world's nations and then among its peoples and persons in an equitable manner, taking into account under the group's budget those emissions that are the product of collective activities rather than attributable to individual GHG-emitting acts. Mitigation costs follow directly from gaps between current emissions and these caps, whether for nations or persons. Bigger greenhouse polluters thus must bear larger mitigation burdens, but for reasons entirely distinct from their responsibility for climate change.

Recognizing that it is now too late to avoid environmental damage from climate change, as some such damage has already occurred and more will certainly result from greenhouse gases that have already accumulated in the atmosphere, the human community must take proactive and reactive steps to satisfy or approximate that same primary imperative. Adaptation is inferior to mitigation in that it allows the *prima facie* bad outcome of environmental damage without the worse outcome of climate-related human harm, but becomes essential once that damage becomes unavoidable. Compensation is inferior to adaptation since it allows environmental damage and human harm, but it is a requirement of corrective justice and so forms an essential response to any harm that is not prevented through mitigation or adaptation. Liability for adaptation should be assessed collectively among nation-states in the first stage of a dual-stage procedure, based on their respective contributory fault, since the causes and effects of climate change can only be comprehended in terms of aggregated emissions and both damage and harm across a large territory. In the second stage, nations should assign their collective fault among their citizens on the basis of differential individual fault, accounting also for irreducibly collective enterprises like national defense that are shared by all.

Nothing in the assignment of collective liability for climate change to nations or peoples or the payment of collective damages to the same precludes the internal distribution of these costs and benefits according to the same principles that are sketched above. Indeed, the same considerations of responsibility and justice would require such an internal distribution. The mere fact that the United States is liable for x in adaptation and compensation costs does not entail that each American is liable for exactly $x/311,966,066$.[35] Those with more post-1990 luxury emissions must bear proportionally greater shares of that national liability total, and those with fewer bear proportionally less. Those just born do not yet bear any liability for climate-related harm, as they cannot yet be faulted for anything. What they do over the course of their lifetimes, in terms of voluntary luxury emission producing acts, will determine how much they will owe for the climate-related harm that the nation collectively causes. Likewise, the funds raised through assessments of liability must be distributed according to principles that emanate from the same conception. Potential victims of allowed climate change are entitled to funds for adaptation assistance to prevent their becoming victimized, or to compensation in the amount of their injury if they are harmed, but not more than this amount. The global annual total raised through assessments of liability must correspond to the global total needed to perform these adaptation and compensation objectives, after a one-time liability assessment based on luxury emissions since 1990 is made. This one-time assessment should be used to initiate the remedial fund and establish its institutional support structure, pay for current climate-related adaptation and existing compensation claims and then be targeted toward the development of capacities and technologies that will hopefully one day make its continued existence unnecessary.

This dual-stage formula of using collective liability in assigning nation-state liability and then making internal individual liability assessments to allocate this national liability among each nation's resident population preserves the philosophical advantages of an individualist normative framework while acknowledging the fundamentally aggregative nature of climate-related harm and the social and collective nature of some of its primary causes (public policies, social norms, collective effort toward sustainability and so on). It therefore need not assign to some nascent global climate regime the impossible task of making many billions of separate personal liability assessments, as proponents of a single-stage individualist framework imply.

But it contains an additional advantage, which may be its most important feature: it binds together residents of nation-states in relationships of solidarity and creates incentives for them to work together toward ecological sustainability by linking their fate through a kind of collective responsibility. As Walzer has observed in describing responsibility for unjust war, 'citizenship is a common destiny' and democracy is 'a way of distributing responsibility'.[36] Insofar as liability for climate change is assessed individually in the first instance, citizens would have no incentive for taking the sort of political responsibility that Young describes to work together toward cooperative social and public policy solutions. They would be made to feel responsible to each other, in the way that it is only possible if their personal costs and benefits are tied to those of others. The dual-stage liability allocation model makes it rational for all persons to work towards reducing their own personal contributions toward climate change *and* reducing those of their nation of residence. It encourages nation-states to work together with each other to make what Nigel Dower calls in his chapter the necessary 'cosmopolitical changes' to realize climate justice, as the total liability to be allocated among nation-states depends on the success of all in promoting sustainability. In so doing, it fosters a sense of responsibility that is prospective, looking forward to a more just and sustainable future, and not merely one that is retrospective, looking backward to ensure that past harm is compensated and past environmental damage is controlled, and reminds us that we are all citizens of the same finite planet, bound together in relationships of interdependence and mutual responsibility.

NOTES

1. This chapter has drawn upon material from within Steve Vanderheiden, 'Globalizing Responsibility for Climate Change', *Ethics & International Affairs* 25, no. 1 (Spring

2011): 1–20, © Carnegie Council for Ethics in International Affairs, published by Cambridge University Press, reproduced with permission.

2. As Stephen Gardiner notes in his overview of the research area, the 'core issue concerning global warming is that of how to allocate the costs and benefits of greenhouse gas emissions and abatement'. Gardiner, 'Ethics and Global Climate Change', *Ethics* 114 (2004): 555–600, p. 578. And controversy around financing issues, whether in terms of developed country assistance for developing country mitigation and adaptation activities or in terms of domestic mitigation actions, were largely responsible for the failure at COP-15 in Copenhagen to reach any legally binding mitigation or adaptation protocol.
3. For a more extensive treatment of equity and responsibility in climate change, see Steve Vanderheiden, *Atmospheric Justice: A Political Theory of Climate Change* (New York: Oxford University Press, 2008).
4. As I shall argue below, both turn on judgments of fault-based responsibility for climate change, which is a function of post-1990 luxury emissions. The common normative basis for adaptation and compensation obligations can thus be usefully contrasted with those for mitigation, which are based in equity.
5. Here, I assume that harm prevention is always morally preferable to allowing for avoidable harm and then attempting to compensate its victims.
6. The draft version of the Copenhagen Accord omits any reference to legally-binding mitigation actions and calls for Annex I parties to jointly commit US$30 billion toward finance of developing country mitigation and adaptation efforts by 2012 and US$100 billion by 2020, without any specification for how these amounts are to be assigned among developed country parties to the convention. See United Nations, *Copenhagen Accord* (draft decision), 18 December 2009. Available online at http://unfccc.int/resource/docs/2009/cop15/eng/l07.pdf (accessed 2 February 2011).
7. The Stern Review estimates the cost of stabilizing emissions at 550 ppm by 2050 to cost approximately 1 percent of gross world product, but also estimates that costs associated with unabated climate change to exceed this amount, resulting in significant net benefits from strong mitigation actions. See Nicholas Stern, *The Economics of Climate Change* (New York: Cambridge University Press, 2006).
8. See Martin Parry, Nigel Arnell, Pam Berry, David Dodman, Samuel Fankhauser, Chris Hope, Sari Kovats, Robert Nicholls, David Sattherwaite, Richard Tiffin and Tim Wheeler, *Assessing the Costs of Adaptation to Climate Change: A Critique of UNFCCC Estimates* (London: International Institute for Environment and Development, 2009). Perry, who chaired the IPCC's working group on impacts, vulnerability and adaptation, estimates that full adaptation costs will be two to three times higher than UNFCCC estimates once the full range of climate-related impacts are taken into account.
9. Simon Caney, 'Cosmopolitan Justice, Responsibility, and Global Climate Change', *Leiden Journal of International Law* 18 (2005): 766.
10. The Intergovernmental Panel on Climate Change (IPCC) predicts of unmitigated climate change that 'the impacts of climate change will fall disproportionately upon developing countries and poor persons within all countries, and thereby exacerbate inequities in health status and access to adequate food, clean water, and other resources.' IPCC, *Climate Change 2001: A Synthesis Report. A Contribution of Working Groups I, II, and III to the Third Assessment Report of the IPCC*, R.T. Watson (ed.) and the Core Writing Team (Cambridge University Press, 2001), p. 12.
11. In a recent peer commentary, I attempted a much briefer version of this same distinction between mitigation and adaptation imperatives, but have since realized that a fuller exposition of these distinctions of actions and guiding principles is needed. See Steve Vanderheiden, 'Distinguishing Mitigation and Adaptation', *Ethics, Place and Environment* 12, no. 3 (October 2009): 283–86.
12. Steve Vanderheiden, *Atmospheric Justice: A Political Theory of Climate Change* (2008), esp. Chapter 7.
13. For a rights-based analysis of these justice obligations, see Steve Vanderheiden, 'Climate Change, Environmental Rights, and Emissions Shares', in *Political Theory*

and *Global Climate Change*, Steve Vanderheiden (ed.) (Cambridge, MA: The MIT Press, 2008), pp. 43–66.

14. For examples of this approach, see Peter Singer, *One World: The Ethics of Globalization* (New Haven, CT: Yale University Press, 2002) and Henry Shue, 'Global Environment and International Inequality', *International Affairs* 75, no. 3 (1999): 531–45.
15. Steve Vanderheiden, *Atmospheric Justice: A Political Theory of Climate Change* (2008), Chapter 7.
16. John Rawls, *A Theory of Justice* (Cambridge, MA: Harvard University Press, 1971), p. 72.
17. G.A. Cohen, 'On the Currency of Egalitarian Justice', *Ethics* 99 (1989): 906–44.
18. Ronald Dworkin, 'What is Equality? Part 2: Equality of Resources', *Philosophy and Public Affairs* 10 (1981): 283–345.
19. Steve Vanderheiden, *Atmospheric Justice: A Political Theory of Climate Change* (2008), esp. Chapter 7. Although I argue that developing countries should be included under such mandatory caps, I suggest that caps ought initially to be set at levels above what these countries currently emit, so that an international cap-and-trade system might initially be used to promote their sustainable development, and then to converge on an equal per capita emissions standard.
20. Joel Feinberg, *Doing and Deserving* (Princeton, NJ: Princeton University Press, 1970), p. 222.
21. Robin Attfield, 'Mediated Impacts, Climate and the Scope of Ethics', *Journal of Social Philosophy* 40 (2009): 225–36, p. 226.
22. Iris Marion Young, 'Responsibility and Global Justice: A Social Connection Model', *Social Philosophy & Policy* 23:1 (2006): 102–30, p. 118.
23. *Ibid.*, p. 114.
24. *Ibid.*, p. 115.
25. *Ibid.*, p. 125.
26. Iris Marion Young, 'Responsibility and Global Labor Justice', *The Journal of Political Philosophy* 12, no. 4 (December 2004): 365–88, p. 382.
27. Matthew Adler, 'Corrective Justice and Liability for Global Warming', *University of Pennsylvania Law Review* 155 (2007): 1859–67, pp. 1860–61.
28. *Ibid.*, p. 1861.
29. *Ibid.*, p. 1867.
30. Brian Barry, 'Sustainability and Intergenerational Justice', in *Fairness and Futurity*, Andrew Dobson (ed.) (New York: Oxford University Press, 1999), pp. 93–117, p. 97.
31. Caney (2005), p. 751.
32. Peter Singer, 'Famine Affluence, and Morality', *Philosophy and Public Affairs* 1 (1972): 229–43.
33. Joel Feinberg, 'Collective Responsibility', *The Journal of Philosophy* 65 (1968): 674–88, p. 683.
34. Andrew Dobson, 'Thick Cosmopolitanism', *Political Studies* 54: 165–84, p. 172.
35. The approximate US population when I last checked the US Census Bureau's real time population estimator on 30 October 2010. See http://www.census.gov/population/www/popclockus.html (accessed 2 February 2011).
36. Michael Walzer, *Just and Unjust Wars* (New York: Basic Books, 1977), p. 297.

3. Climate change and the cosmopolitan responsibility of individuals: policy vanguards

Nigel Dower

INTRODUCTION

In this chapter I explore a cosmopolitan basis for saying that individuals whose lifestyles are carbon-intensive have an obligation to play their part in reducing their own carbon emissions and those of others. I shall argue for a middle position. That is, it is neither the case that individuals have no (significant) obligation to do this, nor the case that our obligations are relentlessly overwhelming. The kinds of activities we are obliged to do are wide-ranging, varying from reducing our own carbon footprints to compensatory actions, advocating changes in others and engaging in political action. Furthermore, it is part of a cosmopolitan perspective that this obligation is to be accepted prior to and independent of this obligation being reinforced by international agreements, national laws and regulations, economic incentives or general social pressure to act in these ways. This is important partly because it is only if some people recognize their obligations in advance of wider acceptance that the necessary 'vanguard action' to stimulate essential wider change is possible. I present a cosmopolitan account in terms of the duties of 'non-maleficence' (we ought not to harm others or cause harm to others) and beneficence, but my main interest is in exploring the implications of adopting a cosmopolitan point of view rather than establishing this as a better starting point than other cosmopolitan principles. These implications are significant for any account of serious cosmopolitan responsibility.

I shall take for granted what appears to be an overwhelming consensus that if mean temperature is to be stabilized at no more than 2 degrees Celsius above pre-industrial levels, people in the rich countries at least need to reduce carbon dioxide and other greenhouse (GHG) emissions by 80 percent by 2050. I take it too that it is possible, in a very rough and ready way, to work out what the average level of carbon emissions would be for every individual, in terms of their direct and indirect carbon footprint, in order to achieve stabilization. It is very rough and ready because there are so many variables to do with population growth (which itself is something modifiable by human decision), how one apportions indirect carbon footprints and so on.[1] What then should individuals do about this?

COSMOPOLITAN CONSIDERATIONS

Implicit in most climate change ethics is a cosmopolitan approach. That is, most people concerned about climate change are concerned about the effects of climate change on present and future humans all over the world. It is not the effects on themselves alone or their countrymen alone that concerns them; they are also concerned about the effects globally. In the light of these effects, they accept that we have obligations to reduce our carbon footprints. It is generally accepted that we also have reasons of collective self-interest to do so, but an important dimension of this is the acceptance of global responsibility. One modern expression of cosmopolitanism is given by Thomas Pogge (see the Introduction to this volume). Pogge identified three main theses of cosmopolitanism: individualism, universality and generality. In summary, it is individual humans that primarily matter (rather than collectivities), all human beings matter equally and the responsibility to achieve human wellbeing is shared by all humanity rather than being restricted to within a community or state. The major ethical theories currently put forward – Utilitarianism, Kantianism and Human Rights theory – are cosmopolitan at root. Despite differences, all of these approaches have clear commitments to universality (for example, the notions that everyone counts for one and no one counts for more than one, and that rights belong to humans *qua* humans) and to global responsibility (for example, 'basic rights are everyone's minimum reasonable demand upon the rest of humanity').[2]

Non-maleficence and Beneficence

All three approaches accept (or at least entail) the principle of non-maleficence and the principle of beneficence. The principle of non-maleficence is a strong secondary rule for utilitarianism, only to be overridden when a greater good is to be achieved. It is implied by Kant's categorical imperative as illustrated by Onora O'Neill's principles of non-coercion and non-deception, and it is implied by the almost-universally accepted correlative duty not to violate the rights of others.[3] The principle of beneficence – we ought positively to promote the good of others, including opposing what harms them – is also a feature of all three approaches. In utilitarianism it is a strong secondary rule only to be overridden when one's own good is more crucially affected by one's action than that of others. In Kantian thought the maxim of helping others is one that is to be universalized, and the idea that rational agents should help other rational agents is illustrated in Nagel's Kantian account of other people's reasons for action giving one reason to act in support of them.[4] The duty to promote human rights is generally (though not always) accepted and includes both the duty to promote or sustain the conditions in which rights can be realized and the duty to help those whose rights are not realized.[5]

This thumbnail sketch of some major ethical theories shows how they can plausibly be seen as cosmopolitan, especially in regard to global responsibility, and as supporting both a principle of non-maleficence and a principle of beneficence. I am deliberately interpreting these broadly so as to allow interpretation of harm and good to include accounts of these in terms of theories of justice and rights that, for instance, enter into the understanding of harm. In what follows I am going to examine the nature and extent of individual responsibility in regard to climate change through these two principles. The main argument is that we need to change our actions in order to reduce the extent to which we are involved in harming other people now and in the future through our own and others' carbon-intensive behavior, present and past. The second argument is that we also need to apply some principle of beneficence, and on this basis to accept that we also have a duty of advocacy and public engagement, specifically the duty to promote the good of reducing the negative effects of climate change more generally. In sum, any reasonable cosmopolitan theory of worth includes at least acceptance of these two principles in some form, or of some principle leading to much the same prescriptions. I do think – though I am aware that I am not offering any further argument for this – that the three main cosmopolitan ethical theories I have mentioned, and indeed many others, include acceptance of these. My main interest is in examining the complex implications of *any* moral theory according to which we ought to make radical changes in the way we act. If, for instance, someone prefers to focus on not exceeding our entitlement to a share of the atmospheric commons without linking it to non-maleficence, and also accepts that we currently exceed this entitlement in a very big way, then many of the same issues arise.[6]

Cosmopolitanism and Sustainability

There is one important clarification to the characterization of cosmopolitanism to make here. Pogge speaks of the equal status of 'every living human being'. This needs interpreting (or amending) to mean 'every present and future living human being'. Of course it is not necessary for a cosmopolitan to accept that future generations of human beings matter ethically because there are arguments in environmental ethics that seek to show that future humans do not have any moral purchase on us. I shall not pursue this objection here. Instead, I will assume that future humans have interests and indeed rights that give us reason to act now. What I am giving is a conception of cosmopolitanism for acceptance, not a definition that is acceptable to all cosmopolitans. However, cosmopolitans interested in climate change are generally concerned with its impact on future human beings as well as with those currently alive. In other words, their understanding of the ethical significance of sustainability is that we have duties to sustain the conditions for a tolerable life for humans of the present and the future. The kind of cosmopolitanism I am advocating here is one that includes sustainability, seen specifically as indefinite sustainability. We might call it 'undated cosmopolitanism'.

Cosmopolitanism and Wellbeing

There are issues about exactly why all human individuals matter equally, and about the nature and extent of transboundary responsibility. It is important to think about what most modern cosmopolitans would accept, and to see what this implies for climate change. We can begin by considering the meaning of human wellbeing. Whatever else we may think is involved, several basic minimum conditions are usually assumed: enough food and water available on a sustainable basis; health and access to medical support for standardly treatable illnesses and injuries; adequate shelter and clothing on a sustainable basis; a suitable physical environment that is both healthy and produces a steady supply of goods; sustained security from arbitrary attack; the possession of life skills and capabilities, acquired *inter alia* through adequate education, and all of this sufficient in normal circumstances for persons to achieve these basic goods and to lead fulfilled lives. In addition to this minimum, but still essential, are further conditions that enable human beings to lead the kinds of lives that, in Sen's words, 'they have reason to value'.[7] This at a minimum requires that individuals are not arbitrarily repressed by social or legal restrictions on their liberty and that they have sufficient resources to be able to do worthwhile things, to enable their dependents to do worthwhile things and to have meaningful relations with others – family, friends and community.

This open-ended account of what is central to human wellbeing, at least as relevant to what others should do about it, is consistent with many different specifications of people's conceptions of the good life.[8] This contrasts with earlier conceptions of universal morality, which involved very specific conceptions of human wellbeing in terms of particular religious theologies and moral codes associated with them. That kind of cosmopolitanism can of course lead – and often did in the past – to crusades, missionizing, proselytizing and so on. I have called this 'dogmatic idealism'.[9] In what follows, I shall assume the above liberal account of what human wellbeing involves. The question then concerns, from this cosmopolitan point of view, what needs to be done to enable all human beings to achieve wellbeing in this sense. We should note that it is wellbeing (in all the above dimensions) that matters, not merely environmental security. Any treatment of climate change ethics needs to account for this.

Two Questions for Cosmopolitanism

There are two ways of approaching the issue: one from the point of view of individual responsibility or commitment, the other from the point of view of what a cosmopolitan might recommend should be done and what should be criticized in existing practices. That is, the question of 'what is the nature and extent of our obligations in regard to the conditions of human wellbeing?' needs to contrasted with the question of 'what, given one's cosmopolitan position, ought to be done in the world in order for the conditions of human wellbeing to be realized everywhere, now and in the future?'. The cosmopolitan certainly needs to propose certain social practices, certain policies and institutions in his own country, certain priorities in foreign policy, certain forms of international law and agreements and certain forms of global governance. Because existing practices are, for most cosmopolitans, far from adequate in delivering the goals of cosmopolitanism, the role of the cosmopolitan is that of advocating and critiquing practices, policies and so forth. This relates to many areas of global concern – world poverty, human rights, war and peace and of course environmental degradation – but it is climate change in particular that provides the clearest example of cosmopolitanism as advocacy and critique. However, this chapter is concerned with questions raised earlier: quite apart from what one advocates, what are one's responsibilities as a cosmopolitan in regard to climate change? What should I do here and now?

Cosmopolitanism and Optimism

Whilst the question of what individuals should do is not to be determined by the laws, economic incentives or established cultural norms that are in place, there is nevertheless a minimum precondition that needs to be brought out. The duty that individuals have is to play their part in reducing human-induced carbon emissions (by themselves and by others). Playing one's part implies that there is some grand goal to be part of. This grand goal is to reduce carbon dioxide and other GHG emissions (at least in rich countries) by something like 80 percent by 2050 so as to stabilize global mean temperature at no more than 2 degrees Celsius above pre-industrial levels. This also implies that others need to play their part, too. Many people are doing this, but most are not – or at least they are not doing enough. There has to be a cautious optimism that at least it is *possible* that, sooner or later, major changes will occur that will take us to the goal of temperature stabilization. I am not saying that it does not make sense for someone to accept that she ought to reduce her carbon footprint to some minimal level even though she believes that virtually no one else will do that and that there is no chance that the world will come to its senses.[10] It is rather that the argument I am presenting about the individual's global responsibility is premised on at least the possibility of major positive change. This is at least characteristic of the cosmopolitan approach because it is not merely premised on an ethical imperative but on a relatively optimistic claim about the possibility of global community and about the resources within human nature to build such a community. The responsibility to play one's part is not based on the firm belief that it will all come out right in the end – sadly, far from it – but on a possibility that is only likely to be actualized if, among other things, a large number of people come to accept this responsibility.

THE NATURE AND EXTENT OF MORAL RESPONSIBILITY

If we accept that individuals ought to change their behavior prior to and independent of law or social sanction, how much and in what ways should we do so?

What is Possible for Individuals?

Should I immediately cut my carbon footprint drastically? For many people, if they had the will to do so, they could do this tomorrow. It would involve some rather painful decisions – decisions which most of us with lifestyles that are carbon-intensive are not really prepared to take just like that. There are now ways, for example through websites, for people to take carbon inventories of their own lifestyles and to ascertain how much they have to do to reduce their carbon emissions to a level roughly equal to their share of the benefits of the atmospheric commons. So the question is: to what extent *ought* we to do what we *can* do? A government's freedom to make radical changes is constrained by various considerations, which may be partly in ethical terms, such as duties to electorates, and partly in terms of what is politically possible, such as the constraints of international agreements. So likewise individuals may be limited in their freedom because they are psychologically unable to make more than certain levels of change or they may be morally entitled not even to do all that they could (psychologically) because of other moral considerations. However, even if we grant that people with typical carbon-intensive lifestyles have psychological limits as to how much they *can* do to change, it is implausible to suggest that they can only change at the rate and in the manner that governments realistically can. (The analogy of the difference in turning circle of a large ship and of a small boat is relevant here.)

The question still remains: ought we to do all that we can? I shall argue that because there are so many other moral considerations to be taken into account, there is no straightforward answer to this question. Nevertheless because there are relevant principles here that apply to the general conduct of our lives – because almost everything that we do has implications for carbon emissions – they ought to *frame* our decision-making in always being at least in the background and frequently in the foreground of practical thinking. While it is hard to pin down *precisely* what each of us should do, the key normative point is that we cannot ethically avoid the question of whether we can we justify the kind of life we lead in the face of our carbon footprints.

The Priority of National Policies and International Agreements

Much discussion about climate change ethics is conducted in what may be called an 'internationalist' context. It is about nation-states (which I shall call states), and what states have contributed in the past to emissions, what their contribution is now, what each state should do, what international agreements need to be made, what principles ought to guide these agreements and so on. The rationale is partly cosmopolitan in that at least some of the drive to reach agreements to cut GHG emissions arises from concern for the long-term prospects for living conditions of human beings anywhere. (Of course there are other motives as well, not least concern for the future wellbeing of people within one's own state.) Much attention is devoted to identifying ethical principles, sometimes stated as principles of global justice, for determining what each state should do, and to pragmatic principles that could be the basis for agreements, such as 'contraction and convergence' and 'greenhouse development rights'.[11]

All these issues and their resolution are of great importance. Without increasing consensus on principles sufficient for robust international agreements between states, the major changes that are needed will not happen. But this chapter is concerned with what *individuals* ought to do.[12] This may seem surprising because it may be thought that, without an answer to the big political questions, we cannot answer the personal questions. This should not be accepted for several reasons. First, what individuals should do is not simply a function of what their state should do. If a country should cut its emissions by x percent by date y, it by no means follows that individual A should cut her emissions by x percent by date y. This argument applies everywhere, and thus includes the carbon-intensive lifestyles of individuals living in countries that have much lower average carbon footprints, a point stressed by Harris.[13] A wealthy Kenyan whose carbon footprint is as high as a wealthy Briton cannot hide behind the fact that Kenya's average carbon footprint is much lower than the United Kingdom's.

Second, it is a regrettable consequence of focusing on national politics and international agreements that attention is diverted away from what individuals should do, especially because of assumptions that how countries reach their targets is up to each of them. This might include many assumptions, such as the expectation that individuals need only play their part by following regulations, or by responding to tax and economic incentives or by being 'green' only to the extent that new cultural expectations have come to be established, or the belief that inequalities in carbon emissions between citizens need not concern individuals themselves. Third, behind the rationale for the first two points is the point that a cosmopolitan account of climate change ethics precisely stresses that as 'citizens of the world' we have an individual responsibility in relation to what happens in the world. Cosmopolitanism is not merely about claiming that the wellbeing of all human beings matters equally, and thus what happens to them as a result of human practices and policies needs to be considered (challenging and controversial as that claim still is). It is about what agents are to do as members of a world community. Even the responsibility of states comes back to the individual insofar as the individual can play her part in influencing what happens politically – what we might call the 'advocacy-public engagement' aspect of cosmopolitan responsibility.

The Principle of Non-maleficence

Suppose that we can make sense of what our share of the atmospheric commons is, in the sense that if everyone had their share, stabilization of atmospheric temperature would be achieved and sustained. Suppose too that it would be wrong for someone to exceed her share or, in rights language, to exceed her 'atmospheric entitlement'.[14] On what basis might we say this? A plausible basis for saying this would be to say that we are harming people.[15] This might be understood as causing people's lives to go badly in some significant way or, to express the idea of harm in a particular way, that we are violating their right to their share of the atmospheric commons by having more than we are entitled to. Some account of what makes a life go badly is needed in terms of the basic goods mentioned earlier because the mere fact that someone else's life does not go as well as it might have done had I not done what I did does not make my action wrong. Many actions in life have this competitive character (that is, if I get what I want, you don't get what you want). If, as a result of current carbon emissions, life conditions in the future are generally less good than they might have been, it is not clear that we are harming other people (though we might have other moral reasons for not wanting to be a party to worsening life conditions in the future). However, if extreme poverty, hunger, violence and environmental degradation on a larger scale will be amongst the consequences, then these are unacceptably bad conditions for life.

The basic idea then is that the carbon-intensive lives that vast numbers of people are leading today are going to cause serious degradation of living conditions in the future for very large numbers of people, and indeed on some analyses have already contributed to greater poverty in some parts of the world because of changing weather patterns. Although the negative effects will be felt all over the world, some areas and groups of people will feel the effects disproportionately, for example the flooding of low-lying territories in Bangladesh and the Maldives. On this analysis we can say that almost everyone at least in the richer countries is, by the way they live, living a life that is morally wrong and indeed seriously morally wrong, and continues to be so as long as their carbon footprints exceed their share. However, if we accept that 'ought implies can' then we need rather to say: we ought as far as is possible to reduce the extent to which we harm others. But matters are rather more complicated than this.

IS PERSONAL CARBON REDUCTION ETHICALLY RELEVANT?

First we need to face the challenge of irrelevance. That is, we need to dispose of an objection that stops the argument at first base. This objection is as follows: individual actions to curb one's own emissions make no difference to whether climate change will take place, and, at least for most individuals, whether their whole life is carbon-intensive or carbon non-intensive makes no difference because the former does not cause climate change, nor does the latter stop it. Walter Sinnott-Armstrong has argued this.[16] He does not argue that we should do nothing but rather that we need to engage in political campaigning to get new laws, new international targets agreed and so on. This is presented as a utilitarian argument, but as others have argued, such a way of interpreting utilitarianism is not undisputed.[17] But the point at issue here does not depend on utilitarianism. If my actions (or whole lifestyle) make no difference to whether a negative outcome occurs, it may be said that my actions (or whole lifestyle) do not cause any harm and my radically changing them does no good in not preventing any harm.

One response is to say that though some climatic changes are such that they will not happen unless some threshold is reached, such as significant weather changes (which is the kind of change Sinnott-Armstrong is considering), other changes such as sea levels and average mean temperatures rising seem to be more the consequence of the cumulative effects of a myriad of small acts. Even here however one's contribution is so infinitesimally small, and the major changes are or are not going to happen irrespective of whether one changes one's behavior, that the same conclusion might be drawn: the difference is so minute that, compared with the positive benefits that come to oneself and others around one, there is no duty to cut one's personal emissions, let alone a powerful duty to make serious changes. Just considering one's actions in isolation seems to lead to this conclusion whether or not one is a utilitarian. (Below I consider the possibility that one's whole lifestyle, taken in isolation, could make a *small* difference, but argue that this is not the main basis for our responsibility.[18])

But thinking of my carbon-reducing acts in isolation like this is to seriously misunderstand the nature of moral reasoning. This can be brought out through another more specific argument: if, as is typically the case for those of us who travel by plane, one travels by plane on scheduled flights, then those flights take place whether or not one is on them: either they are full if someone else has one's seat or they are only heavier by the amount of one's weight and that of one's baggage. As a proportion of the plane's total weight this is very small, so the carbon footprint of one's travelling compared with not travelling is very small (or non-existent if one's place is taken by someone else), and so, given the kinds of reason that people who fly have, the negative moral value of this small contribution is overridden by these reasons. But this account of the effect of my flying is not an adequate way of assessing the ethical nature of my decision, even from a utilitarian point of view. One difficulty is over the contested method of calculation. What is the proportion of the plane's weight that is ethically significant? Surely it seems more reasonable to calculate the proportion as the total weight of the plane divided by the actual or average number of passengers on it. Thought of in this way, one's carbon footprint may be quite significant even for one long-haul flight. However, this still is not the real point. The key question one has to ask is why planes fly, and this has to do with the demand of passengers, so it is the average number of passengers that determines the footprint of planes flying at the rate they do.

Consider a parallel example of ethical consuming. Why do many of us prefer to buy 'fairly traded' bananas, coffee or tea? The actual effect of each one of us doing so or not doing so is minimal or almost non-existent, but still we see such activity as morally significant. Apart from the symbolic aspect of not wishing to be beneficiaries of practices that are unjust, we do accept that we are part of large economic causal chains in that the habits of consumers have effects at a distance. The same applies to travelling. If carbon emissions are causing harm, and much of this is linked with consumer demands for readily available flights (or for goods from abroad), then we are contributing in ways that are far more significant than the simple argument we started with. There are two aspects to this greater complexity.

Mediated Responsibility

Attfield identifies a number of ways in which our responsibilities are mediated, and in two important respects this applies to climate change ethics. One is the sense in which one's actions are part of a set of actions the cumulative impact of which is seriously negative: it is not so much the effect of one's own individual acts as the effects of the class of acts of which they are instances.[19] Attfield quotes Parfit in this regard, but in fact the idea had already been elegantly expressed by J.S. Mill when he said that the wrongness of an action was in virtue of its belonging to a class of acts which are 'generally injurious to society'.[20] Mill's example was lying (because the general practice of lying would undermine trust and mutual reliance and thus have bad consequences for people generally), but it equally applies to our frequent carbon-emitting acts. The other form of mediated action is when, although we do not directly do some damage, other people do some damage in actions they would not perform but for the fact that one performs one's own particular actions.[21] Most of the impacts on the environment we are responsible for are cases of other people doing things so that we can do the things we wish to do, whether, for instance, it is generating electricity for our gadgets or transporting food and other goods from great distances for us to consume or use. These senses of mediated responsibility as applied to climate change clearly show how each individual is implicated in his or her general behavior in climate change issues and suggests a robust account of what is required of each individual agent.

We can apply the idea of mediated responsibility to harm, with an account of mediated harm. In order to make sense of the idea of our doing significant harm to other people through our carbon-intensive behavior, we need to extend the idea of harm in two ways. We harm people because people generally doing the kinds of things that we are doing impacts others and will do so more and more in the future. And we harm other people because of the whole pattern of carbon-intensive activities involving the extraction of resources, the production of goods and their transportation across the world. Furthermore, if we accept this sense of harming, we have to acknowledge that our harming did not start yesterday, but goes back a long way because we ourselves have been part of these complex causal chains of carbon emissions. Suppose then that we accept that the typical activities of people in industrialized countries harm other people on a large scale. What is the ethical significance of this? Given the principle of non-maleficence, it might seem straightforward: change your lifestyle, including your dependence on the harming activities of others, so that you no longer harm people or, if that seems unrealistic, change your lifestyle and dependence on the harming activities of others as far as possible. But, as we shall see, for a variety of quite different reasons, the moral reasoning cannot be as simple as this. However, the point remains: if one is mindful of the wider analysis of harming, one has good reason to change one's lifestyle in many different ways.

Unmediated and Mediated Harm

Let us consider what I shall call 'unmediated harming' (violating other's rights). Here we have a case where it is foreseen (or it is reasonably foreseeable) that harm will occur as the direct result of a specific act one may perform. Even here the fact that an action harms others does not settle the question of its moral character. Putting aside cruelty or sadism, and considering intentionally harming someone else as a means to another goal, such an action done for selfish ends or for morally insignificant considerations would be regarded as wrong. However, if it is done for a serious moral end then, in some circumstances, it may be regarded as justifiable – although the justification has to be a matter of considerable moral importance and not merely a matter of a small balance of good outcomes over bad ones. The moral significance of an anticipated harm that is not the means through which actions are done but a consequence of it is partly a function of whether it is foreseen as a definite, probable or possible consequence. There is a significant difference between doing something in which harm or a negative outcome is risked and doing something that definitively has this outcome, even if it is not part of the means to the end. Here the dictum, attributed to Bentham, that 'we intend all that we foresee' seems to have validity. How does this analysis help with the case of climate change harm? Clearly, whatever harms our actions and lifestyles contribute to, they are not intended as means to our ends; they are generally rather distant effects, and indeed generally not thought about much. On the other hand, given what we know already about climate change, the large-scale negative effects are definite. We cannot really pretend they are only possible or probable. If the harms we are partly responsible for are distant but definite, this fact is of moral significance.

Before we consider harm mediated in the two ways already indicated, let us consider the idea that my actions do *in themselves* actually cause harm (also via intermediate causal chains, including the acts of others). If one thought that one's actions themselves did have some marginal negative effect on some people (granted that everything else that happened remained the same), the fact that it is distant in time makes no difference. Although one's actions in themselves do not alter the *overall* way things go in regard to climate change (as we saw earlier), one's actions make some difference to some people, in the same way as one's acting to reduce poverty may mean a few people are less poor, even though one's actions do not alter the general state of world poverty (which might, despite one's actions to reduce it, be getting worse overall). It is unrealistic to suppose that any *one* of a person's carbon-intensive acts – such as a particular journey or acquisition – had such a specific negative effect, but it is more realistic to say this of the general character of one's actions or the carbon dependency of one's lifestyle. A particular activity cannot really be said to have the negative effect taken in isolation. Thus there is no automatic argument for saying of any particular act that it is wrong because it causes harm.

The point we have established already shows that the duty not to cause harm is not linked to decisions about particular acts and thus is different from the standard case of the duty not to harm we considered above. If we turn to mediated harm, the analysis is even more complex. My individual actions, and indeed the general pattern of my actions in my lifestyle, are instances of whole classes of actions and, even more significantly, they are embedded in complex causal systems. This shows the seriously problematic nature of our actions from a moral point of view, and underlines the way that they contribute to harming others. Nevertheless, precisely because these relationships between my actions and the wider patterns and systems are so complex, the sense an individual has of her actual specific contribution to the bad effects is a rather vague one. Indeed, it is a *sense* of one's being involved or implicated, rather than any discernment of what exactly the extent of one's responsibility is, that is significant. What this sense of involvement gives us, if we accept it, is an orientation or attitude towards our lifestyles, not a direct guide as to what actions we ought to perform at any given time to avoid their being contributions to harming people.

TYPES OF COMBATING CLIMATE CHANGE

A further difficulty with trying to produce a simple formal principle to follow is related to the sheer variety of activities that may be associated with playing one's part in furthering the reduction of our collective carbon footprint. Changing one's lifestyle is not simply a unilinear process of reducing one's direct and indirect carbon footprint. Decisions to walk more, bike or use busses sit alongside decisions to insulate lofts, put in solar panels, buy more efficient boilers, plant (or pay for the planting of) trees or whatever. Whereas these activities can all be seen as contributing to reducing one's own carbon footprint and hence not causing harm, we do not have a formal principle that would determine whether one does one kind of thing rather than another kind of thing. Nor, according to the argument that I am developing, is this needed. This would only be a problem if one were trying to know precisely what array of actions maximizes the reduction in one's carbon footprint.

But changing one's lifestyles need not just be about that. It may involve many other kinds of compensatory activities. If one's reason for giving money to a developing world charity is at least partly to do with the fact that climate change already contributes to greater poverty in many parts of the world, then this is another kind of activity. Likewise, if one engages in fair trade or ethical consuming, one of one's reasons for this may well be that one wants to reduce one's dependency on companies that exploit the environment as well as the world's poor. This can still be regarded as reducing the extent to which one is causing harm in a mediated way because of one's involvement in the system that causes the problems. But it appears to be impossible to determine just what disengaging from the system completely would entail, and in any case one cannot un-write the history of one's past involvement in and dependency on that system and the consequent benefits arising from that fact. Nor is it necessary or appropriate to seek such disengagement.

Matters start to become yet more complex when we recognize another major form of playing one's part: where the object is reduction of carbon-emissions by *others*. This is not lifestyle change as such but being or becoming a 'change agent' in seeking to get others to change their behavior, or to get political change, new laws and regulations, new institutional measures and so on in place. This involves a wide spectrum of actions, ranging from direct action (for example, against airport expansion plans), through joining NGO pressure groups, influencing political parties from within to writing letters to MPs, organizing community climate change awareness events and simply talking with friends or praying (if the latter is seen by a believer as a relevant contribution to the transformation of values and attitudes). What mix of activities one chooses (or recommends to others) depends both on one's assessment of relative effectiveness and on one's assessment of what is regarded as ethically acceptable. Several things need to be noted about these kinds of action.

First, many of these activities may involve an agent participating in activities that have a greater carbon footprint than they would otherwise have engaged in. That is, actions taken to reduce one's carbon emissions and actions taken to influence political and institutional change, or generally to spread the moral message, may not always coincide. Active campaigning may include travel and other activities that involve more carbon emissions than would otherwise occur. One needs to recognize what may be called the Al Gore factor (that is, Gore did a lot of air travel to promote his film *Inconvenient Truth*). The fact that these types of activity may sometimes be inconsistent with each other is not necessarily an objection to trying to combine them. Furthermore many of these actions involve much time, energy and resources, and whilst they may or may not of themselves increase carbon emissions, the prioritizing of them may mean less time and energy being devoted to making those changes that are focused on reducing one's own carbon footprint. This leads to a more general point, which also applies to those who campaign for better economic relations with the Third World (or for a more peaceful world): effective engagement in social and political activism may require reliance to a greater extent on the very systems that one sees as morally problematic. But if one disengaged or withdrew, one would reduce one's effectiveness as an agent of change.

Second, it is important to recognize that the questions 'what kinds of action are appropriate to take?' and 'how much should I do?' are quite separate things. Someone who thinks that we really ought to do as much as we can to combat climate change might, if she thought personal emissions cuts were of no or marginal significance, still spend all her time, energy and money on social and political campaigning (consistent with other obligations she might accept). Conversely someone who thinks that the main thing is to change one's own levels of emissions might think that this is a responsibility, but only to the extent that other people around her are doing so as well. (I make this remark because I have noted a tendency for some people to think that if one stresses political action or questions the efficacy of personal carbon reductions, one is let off the hook of actually having to make serous changes to one's lifestyle. The argument is not as easy as this.) Third, the question arises over the motivational basis for one's action. At least for most of us, if we want a future for ourselves and our children in which serious climate change is mitigated, then serious political and international changes are needed. If a crucial causal condition for these changes is certain kinds of actions of electorates, then we also have powerful prudential reasons for such actions.

Fourth, given my argument above, our duty might seem to be reasonably understood as a further application of the principle of non-maleficence. That is, we are reducing our harming by getting others to reduce their harming. However, this seems implausible because their harming is *their* harming. Although we are all involved together in systematic practices, we can nevertheless distinguish each person's involvement as something they are responsible for. This is where the second principle – the principle of beneficence – comes into play: we have a duty (not something that is optional charity) to promote what is good or, in this case, the reduction of what is bad, by getting others to change what they do. As with harms, we can act positively in a mediated way via collective agency. The good that is done comes often from acting with others in parallel or collaboratively.[22]

The duty of beneficence as I present it is not meant to be a minor duty in comparison with the duty of non-maleficence. It is plausible, when one thinks of the duty of non-maleficence and the duty of beneficence as applied to ordinary actions, to regard the former as more binding that the latter. (I say 'more binding' rather than simply 'binding' for the reasons given earlier about exceptions.) There is greater freedom to choose the occasions for beneficent action, and it would rarely be the case that one could justify harming someone simply because one was doing some good for another person. However, when one looks at the application of these two ideas to complex relations involving distant effects, as is the case with climate change and world poverty, the sharp contrast disappears. There is far greater choice, I have argued, over how one reduces one's carbon footprint. There seems no obvious reason why one should have a presumption in favor of reducing one's carbon footprint as opposed to engaging in some form of activism. There seems no good reason to try to find a formula for prioritizing one over the other. Nevertheless, the sources of the duty do seem to be different for the reasons I have indicated.

The form advocacy and engagement take depends to a large extent on the context in which they occur. All actions that are directed at consequences need to include, as part of the context, the moral climate in which they take place. This point incidentally applies as much to deontologists, who believe that one of their duties is to further climate change justice, as it does to consequentialists because any concern about furthering outcomes needs to take into account what works. What other people believe morally and do for moral reasons, and what shared public goals there may be, affect what kinds of actions are most likely to do the most good *vis-à-vis* climate change. It is only because there are effective avenues for political change via civil society engagement that Sinnott-Armstrong's positive recommendation for action makes sense. It is when there is a publicly agreed goal that playing one's part has added value and is not merely a moral gesture. To give one concrete example, if the 2009 Copenhagen climate change conference (see the Introduction to this volume) had come up with some serious, legally binding commitments, then what it would be best to do would be different from what is best in the aftermath of its failure to do so, which includes joining in the constructive expression of the pent up anger and impatience of many people who are bitterly disappointed with the outcome.

There is a specific argument worth adding here, namely that of Dale Jamieson, who argues that my actions in changing my lifestyle have a catalytic role, assuming they are done in a public manner, in encouraging others to make similar changes.[23] This is presented as a utilitarian argument and is a rejoinder to Sinnott-Armstrong's utilitarian argument about the irrelevance of personal carbon-reducing behavior. The point here is that it is only because other people are likely to be positively influenced by one's example rather than turned off by it – perhaps they are halfway there in recognizing the problem but need some encouragement to actually act – that Jamieson's argument about the multiplier effect of personal carbon reductions makes sense. Jamieson's argument leads to three points of wider significance. First, it illustrates that our own behavior can contribute to the creation of the right social and political climate in which political and institutional changes are more likely to take place. Arguably, one's engagement in political activism and public engagement generally is more likely to be effective if one 'walks the talk' and is prepared to make the changes in one's own life that one recommends for others. There may be a strong positive causal relationship between lifestyle changes and effective public engagement. Second, to the extent that one's lifestyle changes are done publically and with the intention that they may have effects on others, then we have examples of the same actions incorporating both carbon-footprint reduction and change agency, and thus there are no conflicts. Third, in terms of my earlier analysis, the same activities can therefore be based on both the principle of non-maleficence and the principle of beneficence.

THE RELEVANCE OF OTHER GLOBAL CONCERNS

The next consideration shows how complex the moral reasoning has to be. As a cosmopolitan accepting that one has global responsibilities in not contributing to harms anywhere and also in playing one's part in making the world better, one's concerns should not merely be focused on climate change but also on world poverty (on better aid and on better economic relations generally), on war and peace issues, on the destruction of the natural environment (in addition to climate change), on human rights and so on. If we wish to have an ethical basis for tackling climate change, it has to be embedded in one's wider view as to what ought to be done in the world on many fronts, both by oneself and by others. That is, this wider view needs to include both prescriptions about wider social, political and international change, as well as judgments about the range of things that one should do oneself (where these analyses are somewhat independent of each other). Put another way, climate change justice has to be a part of a wider view of what global justice is all about, which for many people nowadays includes, for instance, issues of distributive justice in the global economy.[24]

It is almost inconceivable that any cosmopolitan rationale for climate change action could avoid also being a rationale for a whole range of measures on many other fronts. The conditions of human wellbeing are undermined in many different ways. Indeed, if one combines all these dimensions of global responsibility, one could have an immensely complex range of responsibilities. This confirms that there is no simple formula for determining just what we ought to do, including whether anyone should maximally reduce her carbon footprint. Someone might choose to devote her time, energy and resources to combating climate change (which is not the same as maximally reducing her carbon footprint) or to addressing world trade, peace or other issues – or to pursuing all of these concerns. They are all worthwhile. But there are essentially trade-offs of time, energy and resources, and there are sometimes conflicts between different goals, such as, for instance, getting locally sourced food as opposed to getting fairly traded food from distant lands.

THE DUTIES AND RIGHTS OF ORDINARY MORAL LIFE

Granted then that our global responsibilities are very diverse – both in regard to climate change action itself and other global concerns – there is still the question of how far we ought to take them. There is a range of other moral considerations that now need to be taken into consideration. These might be regarded as the rights and duties that make up the texture of ordinary morally responsible living. I do not pretend to give a fully worked out account of these or of how they are fully consistent with a cosmopolitan approach. My point is that in almost any moral theory, cosmopolitan or otherwise, some importance is attached to interpersonal duties, such as telling the truth, keeping promises, respecting property and avoiding acting violently. Importance is also attached to interpersonal relationships (as sources of wellbeing, duties and rights) among members of a family or friends or communities, and also, crucially, to the moral propriety of people leading lives that are themselves worth living (for example, the pursuit of interesting careers and so on).

How does this affect the argument that we have general duties to play our part to combat climate change? The ethical demands of climate change ethics need to be integrated into the normal range of moral considerations that guide our lives; they do not replace them. I have argued for the ethical demand as both a duty to reduce our contribution to harm and a positive duty to help reduce it, but the challenge is equally pertinent for any theory of climate change ethics that postulates substantial duties to change what we do for the sake of combating climate change. It is difficult to know precisely how much time, effort or resources someone ought to devote to climate change or any other global issue relative to, say, reasonable support for particular others and a reasonable life for oneself. The question of balance is a universal challenge. I doubt whether even the most ardent climate change activists are ready to dispense with the ordinary moral concerns of life or to claim that there are no dilemmas involved.

Has my account in the end committed me to some kind of 'do as much as you can' position? Consider Peter Singer's famous principle in respect to world poverty: 'if it is our power to prevent something very bad from happening without sacrificing anything of comparable moral importance, we ought to do it.'[25] The many responses to this have tended to fall into two camps: either to deny the appropriateness of this maximizing principle or to accept it, but to interpret 'comparable moral importance' in such ways as to take into account the range of factors mentioned above and thus to rob it of some of its radical power. My inclination is to say, especially if one factors in all the various global evils we ought morally to be concerned with, including climate change, that the complexities are such that any attempt to formulate an 'as much as you can' principle turns out to be so abstract as not to be useful. So one should simply argue for serious moral commitment to climate change action, world poverty reduction and so on, treat them as strong *prima facie* duties and trust that an honest assessment of our global predicament will lead to robust action, alongside the normal priorities of life, and to significant adjustments in the way we pursue those priorities.

THE ENABLING CONDITIONS FOR SIGNIFICANT CHANGE

In the last part of this chapter I wish to consider this question: why are arguments about the nature and extent of personal responsibility important to tackling climate change, in particular to getting related political change to occur? That is, if we switch the focus of attention from what an individual ought to do to the question of what needs to happen in order that serious political and institutional change will take place, part of the answer to the question involves acceptance of the kinds of moral responsibility argued for above. Here I briefly consider this at two levels: (1) an acceptance of responsibility to reduce one's carbon emissions and (2) an acceptance of the responsibility to engage in civil society activism.

The Importance of Lifestyle Changes

If individuals accept responsibility to reduce their carbon emissions and act on it, then to whatever extent that they do so, steps in the right direction can be taken cumulatively. My main interest is in the acceptance of this by significant numbers of people – although probably a minority – *before* relevant political and institutional changes have happened. If more and more people accept the need for them to make significant reductions in carbon emissions and wish their governments to take appropriate measures (without being activists over the issue), then this makes political change both more likely and more appropriate. However, even if such political and institutional changes have taken place, we still have the need for the acceptance of these responsibilities – hopefully by very large numbers of people (though it is unrealistic to suppose everyone) – to ensure that, whatever changes come into place, people are keen to ensure that they are effective rather than finding ways to avoid playing their part.

Civil Society Engagement

Civil society engagement for bringing about political and institutional change can be seen as a duty of citizenship. Given that this is directed to change that is crucial from a global point of view, it can be seen as the exercise of global citizenship, or more specifically as 'globally oriented citizenship'.[26] We should first recall that the willingness in any campaigner to make changes in his lifestyle is arguably an important element in making that activism effective. If one is prepared to live according to the principles that one thinks ought to be embedded in public policy, then this is an indication of sincerity and personal commitment that is likely to give one's voice more authenticity and weight in whatever one does in the public arena. The key point about this engagement in the public arena is that in so doing – whether it is writing letters to parliamentarians or to newspapers, or joining pressure groups, working within political parties or taking part in demonstrations – one is seeking to exert an influence on public policy which is considerably more than the influence exerted on government through its assessment of public opinion.

There is nothing undemocratic about seeking to have an influence on public policy well in excess of one's status as a citizen, whose voice as a voter is meant to be considered equally in a democratic state. It is an aspect of the importance of active citizenship, what Miller calls 'republican citizenship' or the responsibility to act as a citizen in 'res publica'.[27] Because relatively few people do this, the influence of active citizens is higher than that of most inactive ones. Individuals get involved in political action and join with others as part of a 'vanguard' precisely in order to get the wider public to follow in their footsteps. The importance for political change of both active vanguards trying to change the balance of public opinion and of increasingly large numbers of people who want their governments to make significant changes – and to varying degrees show this in their own lifestyle changes – is reinforced if we consider democracy.

The Democratic Issue

As Attfield noted, the problem of getting countries to take their responsibility seriously is tied up with the fact that governments of countries are meant (especially if they are democracies) to do their citizens' bidding. This democratic issue is one important aspect of a more general problem, which is that in most countries governments are meant constitutionally to give priority to their current nationals. This goes against a robust pursuit of global targets based on the long-term interests of all people, now and in the future.[28] How then do we get around this democratic paradox? From a cosmopolitan point of view, the idea that political communities – large or small – should be governed democratically seems an appropriate thing to claim, but it has the consequence that if in any given democratic polity the majority clearly do not want their governments to pursue a cosmopolitan policy, for instance on climate change policies, then in one sense they ought not do so, even if from the point of view of an individual citizen who is cosmopolitan they clearly ought to. The solution lies in the increasing acceptance in electorates of the cosmopolitan point of view on climate change, world poverty issues and so forth. Now, if governments are going to make really significant and perhaps painful steps to contribute to new policies, perhaps in advance of international consensus or agreement, then this will require electorates that are well informed and generally persuaded that this is what their governments ought to be pursuing, and that those people are willing to accept the consequences of new policies.

This is not likely to happen unless individuals in significant numbers are willing to make judgments about their own responsibilities as moral agents to contribute to the process of changing the public culture *vis-à-vis* climate change. We need clearly to distinguish between conforming to laws or moral norms and acting as ethical vanguards. That is, we need to distinguish between the idea of agents conforming to new laws or to new socially sanctioned mores about acceptable carbon emissions behavior – perhaps many people would now be willing to conform to laws and mores re climate change behavior *if* they become well established – and the idea of agents who are willing to put significant amounts or time, effort and money into campaigning for these changes and making personal lifestyle decisions to cut back on their carbon footprints voluntarily and well ahead of the prevailing norms of behavior.

CONCLUSION

The ethics of climate change should not merely focus on what countries ought to achieve. It should also focus on what individuals ought to do, independent of and prior to any legal compulsion or social sanction to do so. This is for two reasons. First, without this focus, and a consequent change in what individuals do and believe, the changes in countries' policies will not occur on the scale that is necessary. Second, it is the cumulative impact of individuals' behaviors that will make the difference. This applies as much to the behavior of the rapidly increasing number of rich individuals in developing countries as it does to rich individuals in developed countries.[29] Indeed, one of the merits of the so-called Greenhouse Development Rights approach is that it expects contributions from everyone above a certain level of material development. I have argued that whilst the moral demands to adjust our lifestyles in the face of climate change (and demands to address world poverty) are not relentlessly demanding, nevertheless in both cases our moral responsibilities are much more demanding than is commonly supposed. They require of us actions that go well beyond what law or social custom dictate. Thus, the ethics of climate change need to pay as much attention to this question of what *individuals* should do, here and now, as to what targets countries ought to set for themselves in the future. This is partly because the seriousness with which countries set targets and then pursue them is largely a function of what many individuals – you and me – prioritize right now in our own lives.

NOTES

1. Singer estimated it as 1 ton per capita in 2004 (P. Singer, *One World: The Ethics of Globalization*, 2nd ed. (New Haven: Yale University Press, 2004), p. 35). Monbiot reported actual per capita emissions in the UK of 9.5 tons (G. Monbiot, *Heat: How to Stop the Planet Burning* (London: Allen Lane, 2006), p. 21). Singer's estimate is no doubt optimistic, but the point is that the gap between any theoretical calculation of a sustainable share and the actual average in developed countries is so great that different ways of calculating it do not affect the key issue.
2. H. Shue, *Basic Rights: Subsistence, Affluence and US Foreign Policy*, 2nd ed. (Princeton: Princeton University Press, 1996), p. 17.
3. O. O'Neill, *Faces of Hunger* (London: Allen & Unwin, 1986).
4. This is the sense of altruism in T. Nagel, *The Possibility of Altruism* (Oxford: Clarendon Press, 1970).
5. Shue (1996), pp. 52ff.
6. This illustrates my wider ethical project of showing the importance of accepting shared values derived from various different intellectual sources. (See e.g. N. Dower, *World Ethics: the New Agenda*, 2nd ed. (Edinburgh: Edinburgh University Press, 2007), Chapter 5; cf. Sen's recent approach in A. Sen, *The Idea of Justice* (London: Allen Lane, 2009).)
7. A. Sen, *Development as Freedom* (Oxford: Oxford UP, 1999), p. 18.
8. cf. B. Parekh, 'Principles of a Global Ethic', in *Global Ethics and Civil Society*, J. Eade and D. O'Byrne (eds) (Aldershot: Ashgate, 2005), esp. p. 27.
9. Dower (2007), Chapter 5.
10. One could, for instance, accept the dictum '*fiat iustitia, ruat coelum*' ('let justice be done and the heavens fall') as applying to one's relieving one's climate change conscience in an altogether tragic world.
11. See A. Meyer, *Contraction and Convergence: The Global Solution to Climate Change* (Foxhole, Devon, Green Books, 2000); P. Baer, T. Athanasiou, and S. Kartha, *The Right to Development in a Climate Constrained World: the Greenhouse Development Rights Framework* (EcoEquity: www.ecoequity.org/docs/TheGDRsFramework.pdf (accessed 5 February 2011)). For a consideration of various principles, see, e.g., R. Attfield, 'Global Warming, Justice and Future Generations', *Philosophy of Management* 3, no. 1 (2003): 17–23.; D. Brown, *White Paper on the Ethical Dimensions of Climate Change* (Philadelphia: Rock Ethics Institute, 2006); P. Harris, 'Implementing Climate Equity: The Case of Europe', *Journal of Global Ethics* 4, no. 2 (2008): 121–40.
12. See also, e.g., the approaches of C. Kutz, *Complicity: Ethics and Law for a Collective Age* (Cambridge: Cambridge University Press, 2000); P. Harris, *World Ethics and Climate Change* (Edinburgh: Edinburgh University Press, 2010); S. Caney, 'Justice and the distribution of greenhouse gas emissions', *Journal of Global Ethics* 5, no. 2 (2009): 125–46; and Chapters 4 and 9 in this volume.
13. Harris (2010). See also Chapter 9, this volume.
14. Such egalitarian accounts are offered in, e.g., A. Agarwal, and S. Narain, *Global Warming in an Unequal World* (New Delhi: Centre for Science and Environment, 1991); D. Jamieson, 'Adaptation Mitigation and Justice', in *Perspectives on Climate Change:*

Science, Economics, Politics, Ethics, W. Sinnott-Armstrong and R.B. Howarth (eds) (Amsterdam: Elsevier, 2005), pp. 217–48; Singer (2004); P. Singer, 'Ethics and Climate Change: A Commentary on MacCracken, Toman and Gardiner', *Environmental Values* 15 (2006): 415–22. Caney criticizes this egalitarian account on the grounds, *inter alia*, that it treats the idea of an equal share of the atmospheric commons in isolation from all the other interests that we have and which need to figure in an overall theory of justice (Caney (2009), 130). My view is that whilst his detailed criticisms are pertinent if this is treated as a separate strict principle, it still remains a strong intuitive default starting point for moral thought (and my own later argument reaches a similar conclusion via a different route).

15. See also Chapter 2 in this volume and S. Vanderheiden, *Atmospheric Justice: A Political Theory of Climate Change* (Oxford: Oxford University Press, 2008).
16. W. Sinnott-Armstrong, '"It's Not My Fault": Global Warming and Individual Moral Responsibility', in *Perspectives on Climate Change: Science, Economics, Politics, Ethics*, W. Sinnott-Armstrong and R.B. Howarth (eds) (Amsterdam: Elsevier, 2005), p. 291.
17. J. Garvey, *The Ethics of Climate Change* (London: Continuum, 2008), pp. 147–51.
18. Exactly the same problem incidentally applies to the political engagement of any individuals. In almost all normal circumstances, whether or not I campaign for political changes, those political changes will or will not occur anyway. Of course, it may be argued that my contribution to a political campaign is likely to be more significant than my direct contribution to climate change mitigation by personal emissions cuts. But if I simply consider the effect of my political engagement in isolation, I might well draw the conclusion that the probability of my making a difference was so low that I had no obligation to do so if all I was thinking of was reducing climate change harm in the world by my part in it. But if we see it as a contribution to what others do in an organized group or in self-consciously parallel but separate acts, then one's activity is morally worthwhile because the overall activity will or may make a difference.
19. R. Attfield, 'Mediated Responsibilities: Global Warming and the Scope of Ethics', *Journal of Social Philosophy* 40, no. 2 (2009): 225–36.
20. J.S. Mill, *Utilitarianism*, M. Warnock (ed.) (London: Fontana, 1962), p. 270; D. Parfit, *Reasons and Persons* (Oxford: Clarendon Press, 1984), pp. 70, 78–82.
21. Attfield (2009), p. 231.
22. Cf. L.B. Murphy, *Moral Demands in Nonideal Theory* (Oxford: Oxford University Press, 2003).
23. D. Jamieson, 'When Utilitarians Should Be Virtue Theorists', *Utilitas* 19, no. 2 (2007): 182.
24. See Caney (2009).
25. P. Singer, 'Famine, affluence and morality', *Philosophy & Public Affairs* I, no. 1 (1972): 231.
26. B. Parekh, 'Cosmopolitanism and Global Citizenship', *Review of International Studies* 31, no. 2 (2003): 44.
27. D. Miller, 'Bounded Citizenship' in *Cosmopolitan Citizenship*, K. Hutchings and R. Dannreuther (eds) (London: Macmillan, 1998).
28. Attfield (2009), pp. 232–4.
29. See Harris (2010) and Chapter 9, this volume.

4. Individual responsibility and voluntary action on climate change: activating agency

Jennifer Kent[1]

INTRODUCTION

Climate change presents as a 'diabolical' problem and represents the greatest challenge to humanity of this century.[2] According to Gardiner, the problem of climate change is characterized by three key factors: complexity, lack of causality and institutional inadequacy.[3] Each of these contribute to what Gardiner describes as a 'perfect moral storm' as they represent areas of ethical deliberation essential to resolving the climate change problem but for which existing ethical frameworks are inadequate. Gardiner reasons that the complexity and longevity of the climatic impacts of anthropogenic greenhouse gas (GHG) emissions is signified by the extension of climate change obligations both spatially, as a global issue, and temporally, as an intergenerational one.[4] Who should bear the costs and burdens of climate change is therefore unclear as there is no single causal agent that can be identified as being responsible for the problem. Climate change therefore demands an unprecedented level of global cooperation which calls into doubt the adequacy of existing institutions to address the problem. This positions climate change 'as the moral challenge of our generation'[5] and throws up ethical contestations not only internationally but also between each nation and its citizens.

Responses to the climate change challenge remain largely within the province of international institutions that apply 'top-down' strategies to be delivered by states through their national climate policies. However, governments often emphasize responsibility for climate change action at the individual and household level, that is, from the 'bottom-up'. This assumes that the summation of local actions is (or can be) linked up to national efforts which will lead to global changes.[6] How bottom-up approaches, those necessary actions at the local level, translate into global level action has received little attention[7] and is symptomatic of the essential failure of states and their publics to negotiate their respective roles and responsibilities in countering climactic change.[8] The emphasis on climate policy playing out on the international stage has also largely overridden the growing signs of dissent from civil society evident in an expanding grassroots climate movement. This movement displays deep concerns regarding the ability to achieve an effective international agreement with the urgency and social transformation required to deter the threat of catastrophic climate change.[9] Over 5,200 local actions in 181 countries were held on a global day of action[10] in 2009, calling for a safe target of 350 parts per million (ppm)[11] for CO_2 emissions, whereas global negotiations and the majority of nations' target setting remain focused on higher levels (450–500 ppm).[12] This example serves to expose the layers of potential contestation between institutions and civil society and the need for a better understanding of how local and global processes interrelate. It also highlights the prevailing scientific discourse which dominates climate policy, often overshadowing the essential moral character of the climate change challenge.

The aim of this chapter is to call attention to the most local level of action for climate change abatement, the individual, and to assess what factors may create or restrain agency for voluntary action. I propose that there is an inherent emphasis in developed societies on locating responsibility for climate change, both in terms of its causes and effects, with individual actors. The expectation being that, through their 'personal private-sphere' behaviors,[13] actors possess the authority to effectively reduce their greenhouse gas (GHG) emissions. This 'individualization of responsibility'[14] for climate change mitigation lies within the context of a dominant neoliberal discourse that plays throughout the developed world[15] so that the political ideology of individualism now extends into each person's lifestyle choices and behaviors.[16] I will argue however that due to a range of constraints on personal level actions, individual agency is currently significantly thwarted and hampers an ethical response both locally and globally.

RESPONSIBILITY, COSMOPOLITANISM AND CLIMATE CHANGE

Responsibility for climate change confers duties and obligations,[17] but is also a psychological phenomenon which both works at the personal level (as self-control and free will) but also relates at a societal level.[18] So apart from the creation of obligations or duties as described above, it also implies 'ethical and moral values or caring'.[19] Responsibility sits as one of the key doctrines of the global climate regime. The principle of 'common but differentiated responsibility' (CBDR) (UNFCCC, Article 3.1) is contained within both the United Nations Framework Convention on Climate Change (UNFCCC) and its legally binding treaty, the Kyoto Protocol. In this way responsibility becomes an ethical 'more' within the global climate regime, as the CBDR underlines the inequalities between developed and developing nations in terms of their historic draw on the global atmospheric commons, citizen lifestyles and future development pathways. However as Okereke identifies:

> despite the significant moral concepts accommodated in the text of the UNFCCC and the international equity embodied in the Kyoto Protocol, neither of these instruments says anything explicit on the notion of justice that underlies these aspirations and programmes.[20]

In other words, despite its intention the climate treaty fails to provide clear guidance on how an ethical approach, based on rights and responsibilities, would ensure a just global response to the climate change problem and that can be implemented in practice.

Some scholars go further in their critique of the climate governance regime, arguing that it operates to perpetuate existing inequalities. The three economic instruments of the Clean Development Mechanism (or CDM) under the Kyoto Protocol, for example, lock in a type of carbon *colonialization*[21] whereby rich countries of the developed world can offset their 'luxury' emissions in developing nations. Moreover there is growing political pressure on developing nations, especially those with increasing GHG emission trajectories (China and India, for example), to curb their growth.[22] This fails to acknowledge that in a globalized world 'wealthy nations "offshore" the energy-, natural resource- and pollution-intensive stages of production'[23] so that an emissions-linkage is created between the consumerist demands of the 'North' and the cheap labor-intensive production of the 'South'.

Recently normative theories of cosmopolitanism have gained credence in providing an ethical framework under which the intentions of a just solution to global climate change can be formulated.[24] Drawing on the three essential elements of cosmopolitanism of *individualism*, *universality* and *generality*, the following outlines briefly how these might be applied within an ethical frame to the climate change problem.

Firstly, *individualism* places human beings as the central units of concern of a cosmopolitan framework for climate change, so is primarily a matter of social and cultural concern rather than a response to ecological degradation. Secondly, *universality* applies the ethic of equality to each unit of concern, so that each person would have the *right* to an equal share in the global atmospheric commons. Finally, *generality* implies that each individual unit of concern has a moral *responsibility* for everyone, not just some subset: family, fellow citizens or members of their cultural group.[25] Dobson suggests that such cosmopolitanism (which he calls 'thick cosmopolitanism'), can be found in the 'recognition of ourselves in everyone else who occupies the thin skein of humanity of the surface of the globe [and] is the answer to the problem of exclusion and subordination'.[26] So that matters of unequal power where a particular rationality becomes privileged[27] can be addressed through a cosmopolitan obligation which influences the prevailing structural conditions.

Further a 'cosmopolitan obligation'[28] implies that individuals should undertake action on climate change *irrespective* of a state's inaction (for example, where a developed nation fails to act through pursuing progressive climate change mitigation policies). As climate change is a problem of globalization, this obligation would feasibly extend from the local (individual) to the global (collective). The ability of individuals to take action under the conditions of a cosmopolitan obligation to provide a just response to the climate change challenge then comes to the fore. So at this point I turn to consider how individuals could and do undertake action to mitigate the effects of global warming.

INDIVIDUAL RESPONSIBILITY AS AGENCY

> The self is not a passive entity, determined by external forces; in forging their self-identities, no matter how local their specific contexts of action individuals contribute to and directly promote social influences that are global in their consequences and implications.[29]

Taking individual responsibility for climate change implies that actors are able (and willing) to take mitigation actions, that they are *actors with authority*,[30] possessing the power to engage in practices that will effectively reduce carbon emissions. Individual agency in this sense should be distinguished from the 'unintended consequences of everyday activities',[31] such as the 'simple and painless steps'[32] of changing household light bulbs and purchasing energy efficient appliances.

There is also an understanding that 'reflexive' individuals employ 'active agency' which 'connotes the capacity of human beings to reason self consciously, to be self-reflexive and to be self-determining'.[33] 'Active agents' are also bestowed with 'both opportunities and duties'.[34] They create opportunities to take action but also, concomitantly, have a duty that this action 'does not curtail and infringe on the life chances and opportunities of others'.[35] Agency therefore implies a moral duty not only to act but to act without infringing the rights of others, thus expanding the notion of agency set out by Biermann et al. to incorporate a fundamental moral dimension of agency in individual action for climate change abatement.[36] Moreover this moral obligation to act on the part of the individual exists even where governments fail to take action.[37]

The role of agency also needs to be understood as being embedded in an association with structure,[38] so that:

> Modernization involves not only structural change, but a changing relationship between social structures and social agents. When modernization reaches a certain level agents tend to become more individualized, that is, decreasingly constrained by structures. In effect structural change forces social actors to become progressively more free from structure. And for modernization successfully to advance, these agents must release themselves from structural constraint and actively shape the modernization process.[39]

The ability for individual actors to effect social change is thereby contained within the understanding of the agent-structure relationship. Reflexive individuals are not simply conceived as reactive to social conditions but they can also actively intervene to change prevailing structures. There is an acknowledgement, however, that those social actors are both free to act, but that their actions can be curtailed through institutional restraints. Moreover, as Pattberg and Stripple imply,[40] individual action without critical reflection (such as 'small and painless steps') can simply prove to reinforce the prevailing social norm.[41]

VOLUNTARY ACTION AS BEHAVIOR

Voluntary individual and/or household action to reduce carbon emissions is of particular interest to Western governments, as, reticent to prescribe regulatory provisions for their citizens' behaviors and lifestyles, they expect their climate policy objectives (such as GHG emission reduction targets) will be voluntarily fulfilled through personal and household level behavior change.[42] Perhaps, not surprisingly then, the voluntary action that people take around their lifestyles and homes, with particular emphasis on how an individual's behavior is motivated by their concern about climate change, has been the focus of much empirical research.[43]

Whitmarsh describes individual voluntary action as behavior with *intention*.[44] This behavior is understood to sit within a broader range of co-dependent influences (namely, cognition and affect). Voluntary action on climate change focuses on one aspect of this account – the behavioral – but with the understanding that in order to act people need 'to know about climate change in order to be engaged; they also need to care about it, be motivated and able to take action'.[45] This action is dependent on a wide range of influences as individual behavior is a 'product of social and institutional contexts'[46] that create a complexity of motivations and constraints on voluntary action which has received little normative attention in relation to climate change. Whitmarsh further makes the distinction between intention and impact arguing that most research has focused on the *impact* of action (for example, by measuring how much a household's energy costs have been reduced) rather than the *intent*. She captures the relevance of this distinction in three ways: noting that people may undertake actions with the intention of mitigating carbon emissions but that these may consist of 'futile activities', that is, be ineffective; secondly, that intention can reveal the motivations underlying action; and thirdly, intention uncovers the harder-to-conceptualize range of values, beliefs and virtues that underscore pro-environmental behaviors.[47]

Behavioral intention to mitigate climate change draws attention to the academic literature concerned with why people are failing to respond to the climate change threat through changes within their individual lifestyles.[48] There is now widespread agreement that rationalist information deficit approaches (that is, that by providing information about climate change, voluntary changes in behavior will follow) have firstly proven largely defeatist or unsustainable, and secondly fail to acknowledge the complex mix of behaviors, attitudes, values and social norms that undergird behavioral change. 'The widespread lack of public reaction to scientific information regarding climate change'[49] and the 'failure to integrate this knowledge into everyday life or transform it into social action'[50] becomes even more perplexing when placed within the context of people's stated high levels of concern regarding the effects of climate change. At least in the developed world (where substantial impacts are yet to be felt), high levels of concern have been demonstrated along with an acknowledgement that individuals have a responsibility to take action to reduce their carbon emissions.[51]

INDIVIDUAL AGENCY AND THE VALUE–ACTION GAP

Norgaard has noted the disparity between people's concerns regarding climate change and the adoption of low carbon behaviors.[52] The discrepancy between individuals' stated intentions and their actions has been widely described as the 'value–action' gap.[53] There is a range of barriers proposed that contribute to the gap; however, of most relevance here is that people feel they lack the sense of empowerment to undertake actions that will lead to a less carbon-intensive lifestyle.

Empirical research undertaken by Räthzel and Uzzell expose why the value–action gap may be an artifact of the research process itself.[54] Psycho-social research has focused on individual environmental behaviors which, they argue, in turn reinforces individualistic responses. Their argument is based on two core presumptions of individual responsibility and pro-environmental actions. First, that people's concern is primarily focused on problems at the local level and, second, that they possess the power to do something about them. Räthzel and Uzzell found that people display a spatial biasing in relation to their response to issues such as climate change, so that:

> Ironically, then, although people feel that they are responsible for the environment at the local level this is precisely the level at which they perceive minimal problems. The areal level which they perceive has the most serious environmental problems is the areal level about which they feel least personally responsible and powerless to influence or act.[55]

Both the research and responses to action on climate change have remained centered on an individualistic causality and failed to take into account the broader social and political contexts.[56] They argue that people's 'sense of powerlessness might be a reflection of a larger issue, namely the reality of individualization and competitiveness that govern society at large'[57] and that the 'reductionist individualism' evident in a focus on individual level responsibility and action might rightly dislocate people's ability to respond for the good of society as a whole. This 'psycho-social dislocation'[58] is constructed by an artificially created 'dichotomy between individuals and society' and 'the local and the global'.[59] Given the increasing moral complexity of climate change, it becomes harder to imagine how atomized and disempowered individuals will be equipped to respond to climate change on a collective basis, as individualized responsibility shifts from being a reflexive moral imperative to a set of personal lifestyle practices divorced from their social moorings that 'neither sustain [n]or challenge the structuring of criteria for value in society'.[60]

According to some social theorists,[61] the individualization of responsibility is an extension of the modernizing processes themselves. Individuals are therefore both actively engaged in, and responsive to, the conditions of globalization that surround them, down to the very lifestyles they lead. So, where governments and global institutions state that any successful GHG emission mitigation strategy will require significant changes in lifestyles and behaviors[62] '"lifestyle" connotes *individual* responses to/responsibility for social and environmental change'.[63] This has important implications for the role of individual action in meeting climate change imperatives. In determining the efficacy of response, the nature of these voluntary acts, how they are enacted and the relationship between the actions of institutions (whether global, national or local) and individuals becomes critical. It is important then to determine which types of action undertaken at the personal and/or household level will contribute to the best outcome in terms of global environmental change. The following section outlines a preliminary typology of individual action to assist this task.

A TYPOLOGY OF VOLUNTARY ACTION

There are a myriad of ways that individual actors can and do undertake voluntary action to reduce their carbon footprints.[64] I have constructed a typology of voluntary actions (see Table 4.1) which goes a little way in classifying the types of action choices individuals are presented with in contemporary, developed Western societies.

Table 4.1 Three types of voluntary action

Hierarchical *E.g. personal carbon trading*	Individualist *E.g. consumer-based actions*	Egalitarian *E.g. grassroots climate groups*
Compulsory scheme	Voluntary	Voluntary
Transfers responsibility from the state to the individual/household level	Responsibility shifts from 'citizens' to 'consumers'[65]	Responsibility lies with the individual but is also shared with wider society[66]
'Top-down' Power remains with the state and/or global institutions	'Top-down' and 'bottom- up' Two potential avenues of power are revealed: a) state power remains dominant[67] b) state power is 'hollowed out', authority lies with consumers and global organizations[68]	'Bottom-up' Power is shared amongst citizens

This typology draws on Douglas's cultural theory[69] which has been influential in classifying behavioral worldviews on climate change.[70] Cultural theory sets out four distinct profiles that describe people's different views of nature and society: hierarchical, egalitarian, individualist and fatalist. Each discourse expresses different concepts of responsibility and thereby provides a means to expose and track constructs of responsibility within contemporary climate change debate. Fatalists perceive nature as a lottery and climate change outcomes as a function of chance (consequently, fatalists do not engage in climate policy discussions nor do they believe that their individual actions will effect change); individualists perceive nature as resilient and rely on markets to respond to climate change 'stimuli'; hierarchists perceive nature as manageable and prefer the use of regulation and technologically-based 'solutions'; and egalitarians perceive nature as fragile and regard the engagement of deliberative processes and civil society as critical in a climate change response.[71]

The typology attempts to offer a distinction between the types of voluntary actions available to actors based on their cultural preferences. In Table 4.1 I represent these according to the cultural theory classifications of hierarchical, individualist and egalitarian (it is presumed that fatalists don't engage in voluntary action). Contrary to how these preferences are delineated here, none of these three typologies implies a clear cut scope of action; rather (even though people favor a particular cultural worldview) their behavior incorporates characteristics across all three domains. A brief outline of each typology follows.

In a top-down hierarchical approach to climate change mitigation, global agreements are incorporated into national policy which could be prescribed to the individual through compulsory personal carbon trading. Personal Carbon Allowances (PCAs) have been a focus of research and policy deliberation in the UK, where the government has considered a compulsory scheme where individual and household level carbon emissions would be budgeted to fulfill national targets. In brief, a PCA scheme would operate in a similar way to an emissions cap and trade scheme, that is, a cap or limit is initially established and carbon trading on an individual level can occur up to the limit of the cap.[72] The cap is gradually reduced so that the total amount of carbon allowed to be emitted is reduced over time. Individuals would have something like a carbon credit card to 'swipe' to surrender their allowances from their carbon allowance accounts.[73] The principle of PCAs has been found appealing[74] if not practical from an administrative perspective.[75] Voluntary community-based schemes have gained some traction with Carbon Rationing Action Groups (CRAGs) established in the UK, USA, Canada, Australia and recently in China.[76]

Consumer-based actions have been widely critiqued in relation to pro-environmental behaviors, particularly climate change.[77] Voluntary consumer actions range widely from buying carbon offsets, for example, to offset a lifestyle choice such as an overseas holiday, to paying a premium to encourage renewable energy uptake (such as Greenpower);[78] to investing in less energy intensive appliances (from washing machines to solar panels).

Voluntary actions that fall within the egalitarian typology involve engagement with civil society. Again these range in extent from, for example, participating in collective online advocacy (such as Get Up)[79] to taking part in voluntary activities through membership of an environmental organization or a climate action group.[80]

Critical to this discussion is the role of individualistic responses to climate change abatement which fall within the purview of consumer-based action. According to my argument thus far, governments and other institutions emphasize voluntary individualistic forms of responsibility for climate change mitigation. Individuals, however, in perceiving the complexity and extent of the climate threat and sensing their lack of power to enact global level change, instead apply their agency through personal private sphere behaviors.

This leads to two potential pathways for individualistic action. The first pathway, critiqued by authors such as Scerri and Maniates, positions consumer-based action as responding to the prevailing forces of economic rationalism. In their critique the only pathway currently open to actors for pro-environmental behavior is through their consumer acts.[81] However this action, whilst appearing to empower actors within their personal spheres of authority (their homes and lifestyles), diverts individual attention away from challenging the 'knotty issues of consumption, consumerism, power and responsibility'.[82] Individualization for Maniates is symbolic of the wholesale decline in public engagement in democratic processes in the West which can only be 'remade through collective citizen action as opposed to individual consumer behavior'.[83] In the same way Scerri argues that personal actions deflect individuals from considering how these practices shared in common with other members of society have the potential to challenge or support societal values,[84] so that 'personal acts of consumption stand in for citizen's ethico-political commitments. In the place of engaging in a regulating body-politic, individual citizens are called upon to take initiatives and shoulder responsibilities themselves.'[85]

Contrasting the view that the 'individualization of responsibility', endemic in 'Western culture and ideology',[86] is a disempowering force that funnels human behavior down an economic development path, Spaargaren and Mol argue instead that individualization leads to three forms of 'citizen-consumer' power typified by ecological citizenship, political consumerism (for example choosing fair trade products) and 'lifestyle politics'. They define 'lifestyle politics' as 'primarily about civil-society actors and dynamics beyond state and market' and 'about private, personal and individual morals, commitments and responsibilities'.[87] They argue that the demise of the state allows the 'citizen-consumer' to have an emerging role in environmental politics as connections are forged with global-level institutions and processes through consumer practice. This conception of an empowered consumer base incorporates much from the egalitarian typology and opens the possibility for incorporating forms of consumer practice within egalitarian citizen action (one could think of consumer boycotts, for example). Consumerism for Spaargaren and Moll becomes an entry point for greater democratic involvement at both local and global scales (as state power is 'hollowed out' through the modernizing progression of globalization); however, in saying this, they also delineate the form of individualism displayed in lifestyle politics as being distinct from the neoliberalist interpretation provided by Scerri and Maniates:

> [L]ifestyle politics do not favour automatically or exclusively 'individualist' notions of politics and consumer-empowerment. They are 'individualist' policies in a very, specific, circumscribed way. The concept of lifestyle as it is used by Giddens (1991) refers to the cluster of habits and storylines that result from an individual's participation in a set of everyday life routines they share with others. Every citizen-consumer can be characterized by his or her unique combination of shared practices, the level of integration of these practices, and the storylines he or she connects to these practices. *Lifestyle politics then refer to the ways in which individuals at some points in time (especially when confronted with sudden changes, challenges or fatal moments) reflect on their everyday life.*[88]

The relationship between these typologies of individual action – hierarchical, individualist and egalitarian – and the dominant discourse within the climate regime are thereby revealed. Under current arrangements, the climate regime is principally framed in terms of scientific knowledge (consistent with the hierarchical typology) and market-based mechanisms (individualistic typology) whereas the entrenched inequalities between the developed 'North' and the developing 'South' challenge the attainment of global justice (egalitarian typology). Applying this analysis, individualistic notions of responsibility for climate change thereby serve to support and extend existing 'conceptions of justice [that] have a close fit with the dominant neoliberal economic system which actually underwrite the core policies of this all-important regime'[89] and this in turn determines policy direction and programs.[90] The existing climate governance regime therefore reflects 'the wider discourse of international distributive justice'[91] and exposes the essential inequitable divisions between developed and developing nations in terms of rights to the global atmospheric commons and responsibility for historic emissions. Likewise, economic rather than human rights norms[92] come to shape the climate governance agenda and this is further enabled through complicit individualistic action.

INDIVIDUALIZATION OF RESPONSIBILITY AND AN ETHIC OF COSMOPOLITANISM

Given these differing interpretations of individualistic action, how might a cosmopolitan response to climate change be formulated? The rise in the individualization of responsibility for reducing environmental harms through voluntary actions has been charted in the emergence of re/localization initiatives: community-based grass roots 'movements' such as local currency schemes, community food co-operatives and carbon reduction schemes which rely on individuals and households adopting obligations for reducing their environmental impact.[93] Recent academic interest in individual and local responses to climate change has focused on the rise of the Transition Towns movement.[94] Transition Towns are now well established within the UK and have been taken up in other Western democracies such as the USA and Australia.[95] They aim to respond at a local community, district or town level to the issues of peak oil and climate change.[96] However there are many local scale carbon reduction initiatives[97] few of which have been the subject of empirical research. Paterson and Stripple[98] describe and compare a series of local level programs aimed at climate change mitigation through varying forms of carbon accounting (carbon footprinting and carbon dieting for example) as exemplars of the adoption of individual responsibility for greenhouse gas emission reduction. They include perhaps the most prominent individual responsibility model, Personal Carbon Allowances (or Personal Carbon Trading) which would require personal emissions to be accounted for, capped (and potentially traded) and progressively reduced in an effort to curb consumption-based emissions and meet national reduction targets.

Drawing on Foucault's theory of governmentality from which they derive the notion of the 'conduct of carbon conduct'[99] they argue that concerned individuals 'govern' their emissions as 'counters, displacers, dieters, communitarians or citizens'.[100] They describe the 'conduct of carbon conduct' as enabled through government via forms of knowledge (such as personal carbon calculators); technology (such as translating carbon emission into tradeable commodities); and a certain ethic (for example, a low-carbon intensive lifestyle).[101] Five different types of 'carbon conducts' are examined which are gaining some popularity largely within more affluent western nations: carbon footprinting, carbon offsetting, carbon dieting, carbon rationing action groups (CRAGs) and personal carbon allowances (PCAs). Common to these conducts, according to the authors, is that each contributes to the development of 'reflexive subjects' whose private sphere actions contribute to a 'public good'.[102] They conceive individualism of this type as 'responsible agency',[103] which, I will argue, may not be 'democratic agency'[104] nor engender an individual ethic of cosmopolitanism.

Whilst such conducts may prove popular in the longer term, they currently remain favoured by a specific and limited clique within western societies. Seyfang and Hazeltine (2010), for example, describe Transition Towns as a niche transition consisting of higher than average membership of women in the age group 45–64 with especially high levels of education but lower than average incomes.[105] A group that they describe as '"post-materialists" who eschew high-status jobs and consumption in favor of personal fulfillment and (in particular environmental) activism'.[106] Their research establishes that as a movement Transition Towns has some way to move beyond the niche because whilst there has been dispersion across a greater number of local communities, the movement engages only with a very limited demographic cohort, has faced difficulty in moving beyond information delivery to specific programmatic actions, lacks diversity of stakeholders and the building of broader networks within many communities and, though challenging the current dominant regime, has failed as yet to be mainstreamed.[107]

We are therefore alerted to an important caveat to the progress of an individual ethic of cosmopolitanism. If in the governing of carbon conduct individual actions serve to extend, rather than challenge, the existing power structures based on neoliberalist ideals, these embodied social practices both shape and are shaped by the dominant elite. A 'cosmopolitan vision' requires that individuals release themselves from such structural constraints presumably based on the ethical principles on which cosmopolitanism relies: first and foremost, existing embedded inequalities (between nations and peoples) need to be addressed for the cosmopolitan project to progress. These niche behaviors, currently taken up by a small subset of citizens, narrowly defined demographically and largely confined to wealthy developed nations, will need to evolve into wider community mobilization, to create the global reductions in GHG emissions required to prevent the worst effects of a warming climate.

WHAT CONSTRAINS INDIVIDUAL AGENCY?

The above section outlines some of the ways that individuals can act in order to reduce their greenhouse impact, as an agent of the state, an economic agent or a moral agent, and considers the individual in the context of the prevailing techno-scientific and market-oriented discourse around climate change. Emerging individual and local scale carbon mitigation actions have also been examined for the potential shift to a cosmopolitan response. But in what ways are the conditions for individual agency within modern society being constrained? Here I propose that the inhibition of individual agency for voluntary action on climate change abatement can be demonstrated in three distinct ways and will consider each in turn.

1. *Actors lack authority* (that is they are not empowered to take action). Agency derives from a sense of personal empowerment which becomes the basis from which people are able to take action within their spheres of authority. Norgaard's meta-analysis of psycho-social research on individual action in relation to climate change draws on several lines of empirical evidence to support the supposition that individuals in fact feel disempowered and ineffective.[108] She notes Krosnick et al.'s observation that as there is no easy solution to climate change, people no longer take it seriously.[109] Immerwahr identifies the lack of a sense of efficacy as a barrier to action.[110] Kellstedt states that 'increased levels of information about global warming have a negative effect on concern and sense of personal responsibility',[111] supporting Räthzel and Uzzell's contention that people perceive less responsibility for those matters that are least under their personal control.[112] Actors, in effect, are 'choosing not to choose'[113] to engage with issues such as climate change. The global scale of the problem and the enormous power inequities evident at a personal level (compared to governments and corporations) deluge their ability to see themselves as 'authoritative actors'.[114]
2. *Actors lack trust* in the very institutions (namely, governments) that they turn to for action on issues of global complexity and risk, such as climate change. Whereas governments place confidence in their citizens to respond to the climate crisis through their individual behaviors, the public displace their personal sense of disempowerment through the desire for institutional accountability. In response what emerges is a type of 'organized irresponsibility'[115] where climate change becomes another 'risk' 'for which people and organizations are certainly "responsible" in a sense that they are its authors but where no one is held specifically accountable'.[116] Calls for individual responsibility by governments and other institutions raise issues for the public of institutional trust, capability and duty of care.[117] Not only do people perceive an unacceptable level of action from governments on climate change mitigation but they are also cynical that governments are willing to take action on climate change where it is contrary to governments' or other powerful actors' economic interests.[118] People are also alert to the uneven power relationships that operate between the individual and the state and other institutions.[119]

3. *Actors lack reflexivity*. The essential nature of reflexivity can be portrayed as breaking structural bonds in order to unleash individuals' agency.[120] If, on the other hand, individuals act '*without* questioning the norms of the wider society, the possibilities of change will be constrained by certain norms which are taken for granted',[121] setting up a 'vicious circle' where actors, in conducting their daily lives, reinforce the social norms that in turn 'circumscribe individual choice'.[122] Scerri argues that actors in Western society display their individualism as 'elemental particles of society'[123] whose actions are merely 'an instrument of economic development'.[124] The 'individualization of responsibility'[125] has shifted the emphasis of voluntary pro-environmental behavior to the domain of the consumer. Any ethical considerations are thereby subverted into expressions of green consumerism, what Scerri describes as a type of 'ethics-lite'. The linkages between morality and reasons for acting[126] are severed in this atomistic interpretation as actors no longer reflect on their private sphere behaviors in relation to broader societal values.[127] So in the same way as Räthzel and Uzzell propose a 'psycho-social dislocation',[128] Scerri argues that individualization creates a politico-ethical one: 'In the contemporary West, possibilities for achieving sustainability fall foul of a way of life that, while free to exercise sovereign choices over a plethora of opportunities, is increasingly cut-off from political – that is, value- and so power-laden – commitments to inhabiting the ecosphere on ethical terms.'[129]

ACTIVATING AGENCY

Three key constraints have been argued here to the uptake of effective voluntary action on climate change at the individual scale. First, actors in perceiving individual responsibility for climate change abatement, feel disempowered in the face of the complexity and enormity of climate change risk. Second, in acknowledging their essential powerlessness, citizens turn to their governments to take responsibility for climate change mitigation. However governments are seen by their citizens to be equally incapable, ineffective or uncommitted to rise to the climate challenge. Moreover governments increasingly expect that individuals will take voluntary action within their personal lifestyles but outside of a societal contract that sets up the provisions for sharing responsibility – thus creating a sense of distrust. Third, the structural conditions of modernity inhibit the ability for self-reflexive individuals to generate social change because much of their individual action operates to reinforce social norms, or worse, in the absence of reflexivity, the moral bases for voluntary action are subverted through consumerism.

These three constraints are embedded within two 'dislocations': a psycho-social dislocation that creates an artificial dichotomy between the individual and society, and the local and the global resulting in a type of hiatus in action through people 'choosing not to choose'. The second politico-ethical dislocation separates individuals' moral reasoning for taking voluntary action from broader social values. Both dislocations imply the need for deep reflection on the climate change *problematique* at both the personal and societal scale,[130] and suggest the necessity for a re-balancing from individual responsibility to a shared one[131] along with a shift in power from governments and global institutions to civil society.[132]

Moreover these constraints also reveal important repercussions for the way that climate change solutions are constructed between agents and institutions: as a problem of increasing moral complexity[133] situated within a socio-political context of increasing individualization, the individual and collective may diverge rather than converge on action for climate change mitigation. Enacting a cosmopolitan obligation within global climate governance provides one potential counter for this course as it would establish the elements of a common moral platform from which to address the problem of climate change. Such a cosmopolitan obligation would require that individuals 'create collectivities with the relevant capabilities . . . [to form] individual-duty-fulfilling institutions'[134] that confer rights and responsibilities for climate change mitigation at both the local and global scale. These collectivities could act as a foil to the structural constraints on individual agency.

CONCLUSION: BUILDING A SOCIAL CONTRACT ON CLIMATE CHANGE

In considering the nature of such collectivities I suggest two aspects to be progressed. First, there is a need for the development of a 'democratic agency' formulated from a group of individuals with the ability to effectively organize and operate in a democratic manner.[135] Whilst I have identified the problematic and uncertain nature of new niche social movements based on individual and local forms of climate change responsibility, there lies the potential for these groups to become training 'cells' for collective action thereby empowering people, enhancing social learning and political engagement and building political trust. However, as climate change progresses within the context of globalization and the interdependence of nations is further exposed revealed in the mutual dependence on production in the 'South' and consumption of the 'North', the inescapable transnational impacts of climatic change and ultimately interlinking global economic, social and political systems signal the need for generalizing responses. So, operating beyond the concerns of the local and the state, this democratic agency needs to be employed as a 'cosmopolitan citizenship . . . a thicker notion of citizenship involving economic and social support for the victims of uneven economic development and de-industrialization'[136] and which attempts to influence these inequitable structures through political action.[137] Furthermore such a conceptualization of 'we' could feasibly extend these ethical obligations beyond people to incorporate other species and the global environment as a whole, across spatial and temporal scales.[138] So, rather than conceptualizing climate change around polarized worldviews, framed by the discourse of *difference* in the distribution of its causes and effects, principles of cosmopolitanism accentuate the interconnectedness of the lives of all people (*and* people with nature).

The second aspect then relates to how this might be enacted. In acknowledging the need for governments to engage deliberatively with their citizens on emissions reduction efforts targeting behaviors and lifestyles choices, and to overcome the significant constraints to taking voluntary action, the notion of an environmental contract has recently come to the fore[139] – an environmental contract that defines the rights and responsibilities of both state and civil society on climate change and binds governments and their citizens in mutually agreed commitments and reinforcing action.[140] This notion is appealing as it seeks to overcome the underlying issues of political distrust that inhibit individual and collective action on a national scale. However merely 'tweaking' an environmental contract without addressing the underlying legacy within modern societies formed through the processes of modernization and underpinning much of Western governmental policy (such as shifting government responsibilities to the private sector where citizens are often excluded; creating risks that sit beyond control of the nation-state; and generating the conditions where citizens feel powerless and subsequently withdraw from the political process) will probably fail to resolve the challenge of climate change. At the same time others[141] have argued for a similar arrangement at the global scale to enhance cooperation and progress the moral basis for concerted action. Based on the principles of climate justice 'a shared understanding of fairness', a 'fairness "focal point"', would help broker a mutually acceptable agreement between rich and poor nations.[142] The foundation of such an agreement lies, however, in defining a set of mutually agreed principles based on a shared understanding of what is just and fair and resolving the entrenched inequalities between nations.

To address the moral challenge of climate change it is widely accepted that responsibility needs to be shared between states and their citizens. Significant cuts in carbon emissions are required to prevent catastrophic changes to the Earth's climate systems. These cuts will need to come, in particular, from the developed world from changes in individuals' carbon-intensive lifestyles and behaviors. In the absence of prescriptive forms of enforcing personal and household carbon budgets, global treaties will need to be enacted through states and the voluntary actions of their publics. However in recognizing the psycho-social and politico-ethical disjunctures between the interests of individuals and states a new social contract (between nations and their citizens) is required before effective climate change solutions can emerge. There remains one way for this contract to be re-negotiated and that is by individuals '*joining forces with others*'[143] as cosmopolitan citizens to address global inequalities and effect social change.

NOTES

1. This research has been undertaken as part of doctoral research supported through an Australian Postgraduate Award (APA) funded by the Australian Government. This chapter is based on a paper presented to: 'People, Places and the Planet', the 2009 Amsterdam Conference on the Human Dimensions of Global Environmental Change, 2–4 December 2009, Volendam, The Netherlands.
2. R. Garnaut, *The Garnaut Climate Change Review. Final Report* (Port Melbourne: Cambridge University Press, 2008).
3. S.M. Gardiner, 'A perfect moral storm: climate change, intergenerational ethics and the problem of moral corruption', *Environmental Values* 15 (2006): 397–413.
4. *Ibid.*
5. Ban Ki-Moon in United Nations Environment Program (UNEP) 2009, *Climate Change Science Compendium* (September 2009), p. ii.
6. Accountability and Consumers International, *What Assures Consumers on Climate Change? Switching on Citizen Power*, available at http://www.accountability.org/images/content/2/1/211/What%20Assures%20Consumers%20on%20Climate%20Change.pdf (2007) (accessed 4 February 2011); WWF-UK, 'Weathercocks and Signposts, the environment movement at a crossroad' (2008), available at www.wwf.org.uk/strategiesforchange (accessed 29 May 2008).
7. C. Goldspink and R. Kay, 'Systems, structure and agency: a contribution to the theory of social emergence and methods of its study', *Proceedings of the 13th ANZSYS Conference* (Auckland, New Zealand, 2–5 December 2007).
8. K. Bickerstaff and G. Walker, 'Risk, responsibility, and blame: an analysis of vocabularies of motive in air-pollution(ing) discourses', *Environment and Planning A* 34 (2002): 2175–2192.
9. J. Hansen, 'Climate catastrophe', *New Scientist* 195, no. 2614 (28 July 2007): 30–34.
10. See www.350.org (accessed 19 October 2010).
11. According to 350.org current levels of CO_2 in the atmosphere of approximately 390 parts per million (ppm) need to be reduced to 350 ppm based on scientific evidence to avoid dangerous climate change (defined by the IPCC as a greater than 2 degree Celsius rise in atmospheric temperature).

12. Intergovernmental Panel on Climate Change (IPCC), *Climate Change 2007: Synthesis Report. Summary for Policymakers* (IPCC Plenary XXVII, Valencia, Spain, 12–17 November 2007); N. Stern and N.H. Stern, *The Economics of Climate Change: The Stern Review* (Cambridge UK: Cambridge University Press, 2007); Garnaut (2008).
13. P.C. Stern, 'Understanding individuals' environmentally significant behaviour', *Environmental Law Reporter* 35 (2005): 10785–10790.
14. M.F. Maniates, 'Individualization: plant a tree, buy a bike, save the world?' in *Confronting Consumption*, T. Princen, M. Maniates and K. Conca (eds) (Cambridge, MA: MIT, 2002).
15. D. Harvey, 'Neo-Liberalism as Creative Destruction', *Geografiska Annaler* 88B, no. 2 (2006): 145–158; Maniates (2002).
16. M. Matravers, *Responsibility and Justice* (Cambridge, UK: Polity Press, 2007), p. 73.
17. S. Caney, 'Cosmopolitan justice, rights and global climate change', *Canadian Journal of Law and Jurisprudence* XIX, no. 2 (2006): 255–278; P. Singer, *One World; The Ethics of Globalisation* (Melbourne: The Text Publishing Company, 2002); P. Singer, 'Ethics and climate change: a commentary on MacCracken, Toman and Gardiner', *Environmental Values* 15 (2006): 415–422; Bickerstaff and Walker (2002).
18. A.E. Auhagen and H.W. Bierhoff (eds) *Responsibility: The Many Faces of a Social Phenomenon* (London & New York: Routledge, 2000).
19. *Ibid.*, p. 3.
20. C. Okereke, 'Protecting the global atmosphere: The UNFCCC', in *Global Justice and Neoliberal Environmental Governance: Ethics, Sustainable Development and International Co-operation* (London: Routledge, 2008), pp. 99–122, p. 101.
21. *Ibid.* and Douglas Torgerson, 'Expanding the green public sphere: post-Colonial connections', in *Beyond Borders: Environmental Movements and Transnational Politics*, Brian Doherty and Timothy Doyle (eds) (London: Routledge, Taylor & Francis Group, 2008).
22. The Copenhagen Accord (see http://unfccc.int/home/items/5262.php (accessed 19 October 2010)), which was noted by the 2009 Copenhagen Climate Change talks, includes for the first time non-Annex 1 (developing) country emission mitigation actions, though it should be noted that these are currently non-binding.
23. B.C. Parks and J.T. Roberts, 'Climate change, social theory and justice', *Theory, Culture & Society* 27, no. 2–3 (2010): 134–166, p. 139.
24. P.G. Harris, 'Climate change and global citizenship', *Law & Policy* 30, no. 4 (2008): 481–501; A. Dobson, 'Thick cosmopolitanism', *Political Studies* 54 (2006): 165–184; A.V. Saiz, 'Globalisation, cosmopolitanism and ecological citizenship', *Environmental Politics* 14, no. 2 (2005): 163–178.
25. Based on Pogge cited in Dobson (2006), p. 167.
26. Dobson (2006), pp. 168–169.
27. C. Okereke, H. Bulkeley and H. Schroeder, 'Conceptualizing climate governance beyond the international regime', *Global Environmental Politics* 9, no. 1 (2009): 58–78, p. 63.
28. Harris (2008).
29. A. Giddens, *Modernity and Self-identity: Self and Society in the Late Modern Age* (Stanford, CA: Stanford University Press, 1991), p. 2.
30. F. Biermann, M.M. Betsill, J. Gupta, N. Kanie, L. Lebel, D. Liverman, H. Schroeder and B. Siebenhuner, *Earth System Governance: People, Places and the Planet. Science and Implementation Plan of the Earth System Governance Project* (IHDP: The Earth System Governance Project, Bonn, 2009).
31. P. Pattberg and J. Stripple, 'Beyond the public and private divide: remapping transnational climate governance in the 21st century', *International Environmental Agreements: Politics, Law and Economics* 8, no. 4 (2008): 367–388.
32. WWF-UK (2008).
33. D. Held, 'Principles of the cosmopolitan order', in *The Political Philosophy of*

Cosmopolitanism, G. Brock and H. Brighouse (eds) (Cambridge, UK: Cambridge University Press, 2005), pp. 10–27, p. 12.
34. *Ibid.*
35. *Ibid.*, p. 13.
36. Biermann et al. (2009).
37. J. Garvey, *The Ethics of Climate Change* (London: Continuum International Publishing Group, 2008).
38. Biermann et al. (2009); U. Beck, *Risk Society: Towards a New Modernity*, trans. M. Ritter (London: Sage Publications, 1992); Giddens (1991).
39. S. Lash and B. Wynne, 'Introduction', in Beck (1992), p. 2.
40. Pattberg and Stripple (2008).
41. W.J. Gregory, 'Transforming self and society: a "critical appreciation" model', *Systemic Practice and Action Research* 13, no. 4 (2000): 475–501.
42. Examples of climate change information campaigns targeted by governments at individual lifestyle and behavior change include: 'Be Climate Clever: I can do that' in Australia; in the UK, DEFRA's 'Are You Doing Your Bit?'; and the European Commission's 'You Control Climate Change' (see http://ec.europa.eu/clima/sites/campaign/index.htm (accessed 4 February 2011)); I. Lorenzoni, S. Nicholson-Cole and L. Whitmarsh, 'Barriers perceived to engaging with climate change among the UK public and their policy implications', *Global Environmental Change* 17, no. 3–4 (2007): 445–459.
43. K.M. Norgaard, *Cognitive and Behavioural Challenges in Responding to Climate Change: Background Paper to the 2010 World Development Report* (The World Bank, 2009); L. Whitmarsh, 'Behavioural responses to climate change: asymmetry of intentions and impacts', *Journal of Environmental Psychology* 29 (2009): 13–23; K. Bickerstaff, P. Simmons and N. Pidgeon, 'Constructing responsibilities for risk: negotiating citizen-state relationships', *Environment and Planning A* 40 (2008): 1312–1330; Lorenzoni et al. (2007); Lorenzoni and Pidgeon (2006).
44. Whitmarsh (2009).
45. Lorenzoni et al. (2007): 446.
46. *Ibid.*
47. Whitmarsh (2009).
48. Norgaard (2009), p. 14.
49. *Ibid.*, p. 3.
50. *Ibid.*, p. 29.
51. Norgaard (2009); European Commission, in *Attitudes of European Citizens towards the Environment*, ed. D.G. Communication (European Commission), available at http://ec.europa.eu/public_opinion/archives/ebs/ebs_295_sum_en.pdf accessed 30/3/08 (accessed 4 February 2011); N.F .Pidgeon, I. Lorenzoni and W. Poortinga, 'Climate change or nuclear power – No thanks! A quantitative study of public perceptions and risk framing in Britain', *Global Environmental Change* 18 (2008): 69–85; The Climate Institute, *Climate of the Nation: Australian Attitudes to Climate Change and its Solutions*, available at www.climateinstitute.org.au/index.php?option=com_content&task=view&id=43&Itemid=41, (accessed 21 January 2007); Accountability and Consumers International 2007; I. Lorenzoni and N.F. Pidgeon, 'Public views on climate change: European and USA perspectives', *Climatic Change* 77, no. 1/2 (2006): 73–95.
52. Norgaard (2009).
53. A. Darnton, *Driving Public Behaviours for Sustainable Lifestyles*, Report 2 of Desk Research commissioned by COI on behalf of Department of the Environment, Food and Rural Affairs (DEFRA) (2006); P. Macnaghten, 'Embodying the environment in everyday life practices', *The Sociological Review* 51, no. 1 (2003): 63–84; A. Kollmus and J. Agyeman, 'Mind the Gap: why do people act environmentally and what are the barriers to pro-environmental behaviour?' *Environmental Education Research* 8, no. 3 (2002): 239–259; J. Blake, 'Overcoming the "value-action gap" in environmental

policy: tensions between national policy and local experience', *Local Environment* 4, no. 3 (1999): 257–278.

54. N. Räthzel and D. Uzzell, 'Changing relations in global environmental change', *Global Environmental Change* 19, no. 3 (2009): 326–335.
55. *Ibid.*, p. 328.
56. *Ibid.*
57. *Ibid.*, p. 333.
58. *Ibid.*
59. *Ibid.*
60. A. Scerri, 'Paradoxes of increased individuation and public awareness of environmental issues,' *Environmental Politics* 18, no. 4 (2009): 467–485, p. 478.
61. Giddens (1991); Beck (1992); U. Beck and E. Beck-Gernsheim, *Individualization: Institutionalized Individualism and its Social and Political Consequences* (London: Sage Publications Ltd, 2002).
62. IPCC (2007b), p. 12; see also Stern and Stern (2007); Garnaut (2008).
63. D. Evans and W. Abrahamse, 'Beyond rhetoric: the possibilities of and for sustainable lifestyles', *Environmental Politics* 18, no. 4 (2009): 486–502, p. 501 (emphasis in original).
64. Guidance for individuals and households in this matter has seen exponential growth in recent years but to detail this here is well beyond the scope of this discussion. See Accountability and Consumers International 2007 for a comprehensive listing within the UK and USA.
65. Maniates (2002); Spaargaren and Moll (2008); Scerri (2009).
66. Garvey (2008); Harvey (2008); Dobson (2006).
67. Maniates (2002); Scerri (2009).
68. Spaargaren and Moll (2008).
69. Mary Douglas and Aaron Wildavsky, *Risk and Culture: An Essay on the Selection of Technical and Environmental Dangers* (Berkeley: University of California Press, 1982).
70. Michael Thompson and Steven Ney, 'Cultural discourses in the global climate change debate', in Eberhard Jochem, Jayant Sathaye and Daniel Bouille (eds) *Society, Behaviour and Climate Change Mitigation* (Dordrecht: Kluwer, 2000), pp. 65–92; Michael Thompson, 'Consumption, motivation and choice across scale: consequences for selecting target groups', in E. Jochem, D. Bouille and J. Sathaye (eds) 'Society, Behavior, and Climate Change Mitigation', Proceedings of the IPCC Expert Meeting, Karlsruhe, Germany (2000); M. Hulme, *Why We Disagree about Climate Change: Understanding Controversy, Inaction and Opportunity* (Cambridge, UK: Cambridge University Press, 2009).
71. T. O'Riordan and A. Jordan, 'Institutions, climate change and cultural theory: towards a common analytical framework', *Global Environmental Change* 9, Part A: Human & Policy Dimensions (1999): 81–93.
72. G. Seyfang and J. Paavola, 'Inequality and sustainable consumption: bridging the gaps', *Local Environment* 13, no. 8 (2008): 669–684.
73. S. Roberts and J. Thumin, 'A Rough Guide to Individual Carbon Trading: The Ideas, Issues and the Next Steps', report to DEFRA, the UK Department for Environment, Food and Rural Affairs (2006): 4; available at http://www.cse.org.uk/pages/resources/reports-and-publications/12 (accessed 4 February 2011).
74. M.P. Vandenbergh and A.C. Steinemann, 'The carbon-neutral individual', *New York University Law Review* 82 (2007): 1673–1745.
75. C. Lane, B. Harris and S. Roberts, *An Analysis of the Technical Feasibility and Potential Cost of a Personal Carbon Trading Scheme: A Report to the Department for Environment, Food and Rural Affairs*, Accenture, with the Centre for Sustainable Energy (CSE) (London: DEFRA, 2008).
76. See http://www.carbonrationing.org.uk/ (accessed 4 February 2011).
77. Scerri (2009); Accountability and Consumers International (2007); Accountability,

Net Balance Foundation and LRQA, *What Assures Consumers in Australia on Climate Change?: Switching on Citizen Power*. 2008 Update – Australian Survey, available at http://www.netbalance.com/research/WhatAssuresConsumers.pdf (accessed 4 February 2011); Maniates (2002).

78. See www.greenpower.com.au (accessed 4 February 2011). Australian consumers can purchase Greenpower, which is charged at a premium to allow the energy retailer to purchase power from renewable sources.
79. See www.getup.org.au (accessed 4 February 2011). Get Up is an online campaigning and advocacy organization based in Australia with approximately 336,000 online members which campaigns on a range of environmental and social justice issues.
80. There are about 150 local grassroots climate actions groups (CAGs) active throughout Australia.
81. Scerri (2009); Maniates (2002).
82. Maniates (2002), p. 45.
83. *Ibid.*, p. 65.
84. Scerri (2009).
85. *Ibid.*, p. 477.
86. *Ibid.*, p. 469.
87. G. Spaargaren and A.J.P. Mol, 'Greening global consumption: redefining politics and authority', *Global Environmental Change* 18 (2008): 350–359.
88. *Ibid.*, p. 357 (author's emphasis); Anthony Giddens, *Modernity and Self Identity. Self and Society in the Late Modern Age* (Cambridge, MA: Polity Press, 1991).
89. Okereke (2008), p. 99.
90. *Ibid.*
91. *Ibid.*, p. 101.
92. Paterson (2009), p. 103.
93. G. Seyfang, 'Sustainable consumption, the new economics and community currencies: developing new institutions for environmental governance', *Regional Studies* 40, no. 7 (2006): 781–791; G. Seyfang, 'Cultivating carrots and community: local organic food and sustainable consumption', *Environmental values* 16, no. 1 (2006): 105–123; Accountability and Consumers International 2007, *What Assures Consumers on Climate Change? Switching on Citizen Power*, available at http://www.accountability.org/images/content/2/1/211/What%20Assures%20Consumers%20on%20Climate%20Change.pdf (accessed 4 February 2011).
94. G. Seyfang and A. Hazeltine, 'Growing Grassroots Innovations: Exploring the Role of Community-Based Social Movements for Sustainable Energy Transitions', Norwich, UK: CSERGE Working paper EDM 10-08 (2010).
95. G. Seyfang, A. Hazeltine, T. Hargreaves and N. Longhurst, 'Energy and Communities in Transition – Towards a New Research Agenda on Agency and Civil Society in Sustainability Transitions', Norwich, UK: CSERGE Working Paper EDM 10-13(2010): 12. Note for example that there were over 150 Transition Town groups in the UK in Spring 2010 and another 100 other groups worldwide.
96. Seyfang and Hazeltine (2010).
97. The 'What Assures Consumers on Climate Change?' reports (Accountability and Consumers International 2007; Accountability and Net Balance Foundation 2008), for example, list a diverse group of mass awareness raising campaigns and 'communities of change' programs operating in the UK, USA and Australia which focus on individual and household level climate change mitigation actions.
98. M. Paterson and J. Stripple, 'My space: governing individuals' carbon emissions', *Environment and Planning D: Society and Space* 28 (2010): 341–362.
99. *Ibid.*, p. 347.
100. *Ibid.*, p. 342.
101. *Ibid.*, p. 347.
102. *Ibid.*
103. *Ibid.*

104. C. List and M. Koenig-Archibugi, 'Can there be a global demos? An agency-based approach' *Philosophy & Public Affairs* 38, no. 1 (2010): 76–110, p. 89.
105. Seyfang and Hazeltine (2010), 7–8.
106. *Ibid.*
107. *Ibid.*
108. Norgaard (2009).
109. Jon Krosnick, Allyson Holbrook, Laura Lowe and Penny Visser, 'The origins and consequences of democratic citizen's policy agendas: a study of popular concern about global warming' *Climate Change* 77 (2006): 7–43.
110. John Immerwahr, 'Waiting for a Signal: Public Attitudes toward Global Warming, the Environment and Geophysical Research', Public Agenda/American Geophysical Union (1999), available at http://www.policyarchive.org/handle/10207/5662 (accessed 5 February 2011).
111. Paul Kellstedt, Sammy Zahran and Arnold Vedlitz, 'Personal efficacy, the information environment, and attitudes toward global warming and climate change in the United States', *Risk Analysis* 28, no. 1 (2008): 113–126.
112. Räthzel and Uzzell (2009).
113. Macnaghten (2003).
114. Biermann et al. (2009), p. 32.
115. Beck (1992).
116. A. Giddens, 'Risk and responsibility', *The Modern Law Review* 62, no. 1 (1999): 1–10, p. 9.
117. Pidgeon et al. (2008): 75; Bickerstaff et al. (2008); Macnaghten (2003); Bickerstaff and Walker (2002).
118. Darnton (2006): 24.
119. Bickerstaff et al. (2008); Maniates (2002).
120. Gregory (2000); Beck (1992).
121. Gregory (2000), p. 485.
122. *Ibid.*
123. Supiot cited in Scerri (2009), p. 470.
124. *Ibid.*, p. 473.
125. Maniates (2002).
126. Scerri (2009), p. 470.
127. *Ibid.*, p. 478.
128. Räthzel and Uzzell (2009).
129. Scerri (2009), p. 479.
130. Gregory (2000).
131. Scerri (2009).
132. Gregory (2000), p. 499.
133. Gardiner (2006).
134. Jones cited in Dobson (2006), p. 181.
135. List and Koenig-Archibugi (2010), p. 89.
136. A. Linklater, *The Transformation of Political Community* (Cambridge: Polity Press, 1998), p. 203.
137. *Ibid.*, p. 206.
138. K. O'Brien, B. Hayward and F. Berkes, 'Rethinking social contracts: building resilience in a changing climate', *Ecology and Society* 14, no. 2 (2009): 23.
139. S. Hale, 'The new politics of climate change', *Environmental Politics* 19, no. 2 (2010): 255–275.
140. Hale (2010) suggests that the UK's Sustainable Consumption Roundtable report 'I will if you will' proposes such an environmental contract.
141. See Chapter 9 this volume and Parks and Roberts (2010).
142. *Ibid.*, p. 136.
143. Gregory (2000), p. 490.

5. Cosmopolitan solutions 'from below': climate change, international law and the capitalist challenge

Romain Felli

INTRODUCTION

International negotiations on climate change have yet to prove their ability to put the world on the much-needed track toward dramatic reductions in the level of greenhouse gas emissions. Since the Kyoto protocol was devised and implemented, climate change remains as dangerous as ever and the steps towards a real reduction in greenhouse gases emissions have been few. Some may even say that the steps have been mostly backwards. The failure of the Copenhagen Summit in 2009 reinforces such a view. It can therefore be argued that '[t]his failure can be attributed, at least in large part, to the nature of the climate change regime itself, which is premised on negotiations among states seeking to protect or promote their relatively narrow national interests'.[1] This understanding leads to calls for a renewed approach on climate change, based on cosmopolitan ethics rather than on statist positions.

Cosmopolitans tend to oppose state interests to cosmopolitan human interests and seek to make the second prevail over the first. While I generally agree with the cosmopolitan position, I believe that a simple opposition between states and individuals is misleading in understanding the current situation and acting upon it. This apparent opposition needs to be grounded, I will argue, in the capitalist relations of production that suppose the production and reproduction of two separate spheres, that of the 'economy' and that of the 'political'. This separation is enacted at the international level, where the existence of a world market is paralleled with the generalization of political sovereignties, that is, a system of formally equal and autonomous states. International negotiations, and their outcomes as international law, thus seem the result of the expression of the 'narrow national interests of states'. This fetishist understanding of the international realm is itself a necessary expression of the capitalist relations of productions. There is thus a need to go beyond the appearance of 'states interests' in order to reveal the mediations that produce this fetishist understanding. In other words why would state interests appear as opposed to human interests, and what does it tell us for a cosmopolitan perspective?

Therefore, beyond the question of the narrow national interests, and in order to devise a cosmopolitan 'solution' to climate change, we should reflect over international law on climate change, using an historical materialist approach. The analysis will mostly remain at a fairly high level of abstraction, but concretized when necessary. On top of the fetishist inter-state understanding built into the climate regime[2] and expressed in international law, I will argue that two other forms of fetishism are at work. First, a fetishism of distribution, through which capitalism's dependency on fossil energies – hence the sphere of production – is not assessed, leaving only distributional questions open to debate. Second, a fetishism of the international realm, in which international law is seen as a tool for advancing the common good. The identification and analysis of this fetishist understanding is the necessary condition for a renewed cosmopolitanism that would seek to advance human interests regardless of membership. A truly cosmopolitan solution to climate change, I will argue, would involve a transformation in the social relations that are congruous with this fetishist understanding. Some contemporary social movements are actually pointing towards such a critique. The 'Peoples' Climate Summit, KlimaForum09', and the various demonstrations held during the Copenhagen Summit in which tens of thousands of persons took part despite their repression by the police, the 'World People's Conference on Climate Change and the Rights of Mother Earth' hosted by the Bolivian Government in Cochabamba in April 2010, as well as, more generally, the climate justice movement, are contributing to what Fuyuki Kurasawa has called a 'cosmopolitanism from below',[3] a concept with which he supplements contemporary accounts of cosmopolitanism and on which I draw in this chapter.

COSMOPOLITANISM FROM BELOW

Cosmopolitanism is a form of political and moral philosophy according to which the national citizenship of individuals is not relevant in regards to principles of justice. Cosmopolitan theorists affirm the unity of mankind, and the need, despite its fragmentation, to treat equally each and every human being regardless of his or her institutional membership. Or, on a more minimal tone: 'the demands of justice must be decoupled, at least to some degree, from the territorial bounds of the state.'[4] This ethical cosmopolitanism is generally supplemented with a political cosmopolitanism, which supposes a form of global institutional order (short of a global state) through which the egalitarian principles of cosmopolitanism may be enacted, such as a form of global governance.[5] Such a form of governance is particularly advocated in the case of global climate change as it is argued that the problem, while always local in its sources and results, is global in scope: 'Global warming has bound us together worldwide in a morally distinct way.'[6]

The understanding of cosmopolitanism discussed in this chapter will be different in some regards. While retaining the basic assumption of equality between human beings regardless of citizenship, this account will be focused on structures that hinder the cosmopolitan equality of human beings, granted that those structures are not only political orders as mainstream accounts of cosmopolitanism would have it. Rather, we need to understand, first, how social relations of domination exist which are not reducible to (institutional) political domination, and, second, how these social relations of domination are expressed through political institutions, such as international law.

This project also leads us towards a less formal and individualistic understanding of cosmopolitanism than the one used in mainstream cosmopolitan theories, hence the idea of a 'cosmopolitanism from below'. Kurasawa created this concept because of his dissatisfaction with the two major contemporary forms of cosmopolitanism: what he calls normativist cosmopolitans and politico-legal cosmopolitans.[7] These two forms of thought are part of a 'cosmopolitanism from above' which has not 'been sufficiently attentive to the transnationalization of the sources and possibilities for solidaristic action from below, thereby neglecting the processes through which individuals and groups are cultivating relatively thick global social relations'.[8] Indeed, dominant accounts of cosmopolitanism often 'overlook or marginalize a vast array of socio-political action performed by groups and persons who may not seek official sanction or juridical inscription'.[9] They also take for granted or leave 'unexamined systemic factors that underpin socioeconomic and civil-political injustices . . .'.[10] Rather than understanding cosmopolitanism as a purely ethical or normative principle against which the reality of the international system is assessed, I will rather try to locate and analyze social practices of cosmopolitanism through emancipatory struggles against climate change that are enacted from 'below', that is, outside and even against the states.

The standpoint of the cosmopolitan critique of the current climate regime is hence not some form of abstract moral norm, but rather it is constituted by social practices. This cosmopolitanism from below is not necessarily the cosmopolitan civil society of contemporary liberal political theory; rather it is constituted by these social practices, which are necessarily localized and rooted, but which express a concern for – and a solidarity with – fellow human beings across the world, in spite of national boundaries. We should recognize, however, that these practices have not yet reached a point where one could talk of a genuine cosmopolitan alternative to the current climate regime. Nevertheless, and despite their fragmentation and heteroclite nature, these practices reveal the possibility of such an alternative.

INTERNATIONAL LAW AND CARBON IMPERIALISM

In order to assess the fetishisms involved in the climate regime, we need to relate the form of regulation enacted through international law, to basic capitalist relations of production and their political mediations. I cannot develop here a full-fledged account of the relationship of capital to the states-system, but let me briefly sketch its main features. At a high level of abstraction, the form of bourgeois society can be conceptualized as the dialectical 'separation in unity' of the state and civil society.[11] This is coherent with Marx's claim that 'the *establishment of the political state* and the dissolution of civil society into independent *individuals* – whose relation with one another depends on *law*, just as the relations of men in the system of estates and guilds depended on *privilege* – is accomplished by *one and the same act*'.[12]

One characteristic feature of capitalism is thus the apparent separation between the locus of exploitation and that of the reproduction of the relations of productions.[13] The economic sphere is the place where surplus is extracted, apparently without political coercion, whereas the political sphere appears as the place where social relations are reproduced. The function of this extra-economic power is 'the imposition, maintenance and enforcement of social-property relations conducive to the exertion of economic power. . .'.[14] As we know, the abstraction of capital is concretized through the existence of multiple capitals, whose competition allows for the actualization of the law of value. The competition between various capitals, however, remains subordinate to the capital/labor relation. It expresses the separation in unity of various capitals as capital versus labor. Similarly, the concretization of the concept of the state is to be found in the states-system, the political structure that ensures the reproduction of the relations of production. It is expressed in the cooperation and competition of various territorialized states. Sovereignty, as the territorialization of political self-determination, lies at the heart of the capitalist political order.[15] States are, therefore, 'political nodes in the global flow of capital',[16] but the concrete form that these nodes will take remains undetermined at such an abstract level (except for their abstraction from civil society, their plurality and their territoriality). Particular states will take different forms, depending on the transformations of class struggle. Therefore, the nature of the relation between 'the state' and 'the market' will vary, as will the relation between states and their national capitals.

The action of states cannot be taken as functional subordination to the accumulation of capital in general, nor can it be construed as simply reflecting the interests of their national capitals. Rather, they should be conceptualized as struggling in order to attract and retain a share of global capital within their boundaries. This competition is, of course, highly unequal. This process leads to various forms of 'decomposition of global social relations'[17] through which class opposition is replaced by national oppositions. As Nachtwey and Ten Brink affirm:

> the system will only work if the competing members of competing classes – both wage earners and entrepreneurs – are bound together at the state level and are thereby at odds with the corresponding classes outside the territory of the state. Not for nothing is the creation of cross-classes coalitions to safeguard 'national competitiveness' an absolute central neo-liberal argument today.[18]

Put at a more abstract level of analysis, this means that 'the clash of particular and universal, public and private interests is grounded in political struggle over the constitution of what is, in the event, some hegemonic national-popular coalition of interests, as the "universal" public interest'.[19]

The legal form then is the form that the struggle between capitalist states may take for securing the resources necessary for the reproduction of capital within their territories.[20] This struggle is not expressed in an overtly violent manner, but through the mutual recognition of equality. Inasmuch as the legal form supposes free and equal individuals on the domestic stage, it supposes the relations between formally free and equal – that is, sovereign – states on the international level. This is why the concept of 'sovereign equality' is of such vital importance in international law. International law actually is predicated on equal sovereignties: 'State sovereignty and international law are coconstitutive: international law accords the recognition, standing, and rules of behavior for sovereign states; sovereign states are a key source of international law.'[21] Jean L. Cohen, who vindicates this idea from a liberal perspective, specifies that 'formal equality has to be linked to some degree of material equality among the states',[22] which somehow makes her argument circular. Obviously, if 'only equals are equals',[23] there would be no need for a legal form to exist. This circularity is avoided by Miéville, who insists that formal equality in international law goes hand in hand with very substantive inequalities of wealth or power. 'Although both parties are formally equal, they have unequal access to the means of coercion, and are not therefore equally able to determine either the policing or the content of the law.'[24] This accounts for the indeterminacy of international law. However, the differential power between the states is the difference between various *capitalist* states.[25]

Moreover, the legal form constrains the conception of the relation of human beings to nature as a relation of property. Capitalist relations of production suppose a duality in the subject and the object.[26] Nature is experienced as a provider of raw materials that can be used to produce use values, the carriers of value under capitalist relations of production. The development in the legal form is a development in the objectification of nature, especially via the development of property rights.[27] Corrigan and Sayer insist on the role of the law not as a justification or as a 'veil' but as *constituting* these relations by constituting private property: 'Property did not have just to be seized, rather it had to be *constituted*. Legitimation here means more than mere ratification.'[28] Furthermore, 'for the right of property to be *definable* with this simplicity – abstractly and universally, without references to the nature or standing of either owner or object owned – supposes, so Marx argues, definite historical conditions that were long and often violent in the making'.[29] As Tran has argued, the development of capitalist relations of production, in contradictory unity with the development of the modern state, produced the eminent property of the state over the land, which does not preclude private property over portions of the national territory. Actually, the eminent property of the sovereign state over the land is the condition for the constitution of private property of the land.[30]

We can understand, similarly, international law on climate change as constituting a definition of rights of property over the 'earth's carbon-cycling capacity',[31] itself a politically constituted condition of production. These rights are devised and distributed in a highly unequal manner, in line with the relative powers of the capitalist states. This struggle over the appropriation of the earth's carbon-cycling capacity is realized through international law. International law on climate change, however, is not generally perceived as a form of appropriation of the earth's carbon-cycling capacity, but rather as a tool for preserving it and managing the commons. This is due to the fetishism involved in the perception of these regulations. The legal form is inherent in capitalist relations of production. This is why Pashukanis sought to complement Marx's theory of commodity fetishism with a theory of legal fetishism. Fetishism is a concept that seeks to grasp forms of consciousness in capitalism.[32] Although it accounts for biased perceptions, it is not a theory of 'false consciousness'. Fetishism is a narrow and partial perception of a broader set of relations. It is a concept that seeks to account for the adequacy between the kind of relations that subjects maintain with each other in capitalism and the perception that they have of these relations, or, more precisely, of the forms under which these relations appear. For instance, rather than perceiving the commodity form of value as a result of social relations, subjects see commodities as mere objects whose value depends on their intrinsic qualities. The relation between subjects is replaced with a relation between things.

Similarly, law is not perceived as a historically specific social relation congruent with the commodity-form, ultimately based on determined relations of production, but as a neutral (ahistorical) means of coordinating activities in a society made of equal and autonomous subjects. Law is taken as the embodiment of the 'common good' or the 'general will', as proclaimed by a state that stands above civil society, one that is above class struggles. In other words, law as a general, abstract and equal norm is a necessary feature of the capitalist relations of production. These general points allow us to consider the international climate change regime as the competition between various capitalist states (endowed with highly unequal powers) that seek to attract and retain capital within their boundaries, and international law as the form of expression of this relation. As a form of appearance of capitalist relations of production, the legal form is perceived through a fetishist understanding. Taken as an autonomous power that stands above class struggle, and for the international law above national interests, the legal form conceals its origin in the basic capitalist relations of production. In the case of the climate regime, this is translated at three levels, which are being, more or less consciously, challenged by a cosmopolitanism from below which needs to go beyond national divisions and create active solidarity across borders, based on the similarity of class situations and not on national memberships.

THE FETISHISM OF DISTRIBUTION

The climate change regime is predicated on the idea that capitalist economic growth is compatible with (and for some even conducive to) environmental protection, a sustainable use of resources and a decline in pollutions.[33] In this perspective, 'solutions' to the climate crisis are found outside the production sphere: basic relations of production, such as property relations, are not questioned. The result of this understanding is the 'fetishism of distribution'. By this, I mean the systematic framing of the solutions to climate change around the issues of distribution of the rights to emissions.[34] This framing, in turn, allows for the constitution of the solutions of the climate crisis in terms of market instruments. The fetishist grasping of the climate regime is paralleled by mainstream normative political theory which has devoted itself to the elaboration of the criteria by which a distribution of emissions allowances could be labeled 'fair' or 'just',[35] without however examining the relations of production leading to the crisis.

The social relations of production that are at the heart of the global ecological crisis (the 'metabolic rift'),[36] are thus veiled by an understanding of the climate crisis in terms of the regulation of the rights to emissions. A systematic analysis of the basic relations of production in capitalism, in relation to the climate crisis, would go beyond the over-emission of greenhouse gases, and take the extent of the dependence of the capitalist economy on fossil energy as its central enquiry.[37] The trajectory of production under capitalism, which is predicated on the accumulation of value, necessitates an ever-increasing supply of raw materials (as wells as pollution 'sinks') in order to serve for the materialization of use values, thus leading towards a restless path of growth and environmental destruction.[38] Therefore, under the current conditions of production, technological developments and productive forces, a radical departure away from fossil energy as the main source of energy would prove a fatal blow to the economic system.[39]

The fetishism of distribution allows for an understanding of the climate crisis in which the use of fossil energy remains unquestioned, and in which only its side effects (greenhouse gases emissions) are relevant. Political action on climate change, at the domestic or international level, is thus concerned with the distribution of pollution rights, not the control over energy sources or systems of production. Patrick Bond distinguishes four main forms of greenhouse gases mitigations that are currently being discussed: 1) carbon trades without auctions ('grandfathering'); 2) carbon trades with auctions; 3) carbon taxes; 4) per capita 'rights to pollute' strategies, such as 'contraction and convergence' or 'greenhouse development rights'.[40] He adds that there are at least two other solutions that never make the headlines: 'command-and-control prohibitions' and what he calls 'local supply-side strategies' (a kind of command-and-control *from below*).[41] The point is that the first four solutions focus exclusively on the forms of distribution of the emissions rights. They therefore all take for granted the possibility and desirability to define and grant property rights over the earth's carbon-cycling capacity. By so doing, however, they do not address the fundamental relations of production that produce the climate crisis. The form through which this distribution of emission rights is achieved is the market. Property rights and the market thus appear as the best way to tackle greenhouse gases emissions.[42]

This fetishism of distribution is complemented by an understanding in which individuals, generally reduced to 'consumers', are invested with the power to make unconstrained choices in a market offering all sorts of alternatives, and are therefore seen as 'responsible' for the greenhouse gases emission they release through their lifestyles and consumption patterns. This profoundly disempowering trend of depoliticization of the ecological crisis is linked with the rise of 'environmental citizenship'through which an 'emphasis upon atomistic voluntarism and personal-use commodity consumption compromises many of the normative frameworks through which people in the West might act to ameliorate some of the political tensions that the ecological challenge raises'.[43] Here a cosmopolitanism from below would involve a collective, political action, that challenges individualistic and market based solutions to the ecological crisis and asks for stronger collective, conscious, regulations of the relations to nature, be they in the form of a renewed public action (in a social democratic manner) or in more direct actions, such as 'command-and-control from below'. The last two solutions mentioned by Bond deal with the production sphere, by addressing not just the emissions of greenhouse gases, but also the use of the materials that in the process of production will emit greenhouse gases. This focus moves us away from the 'possessive individualism' of the right to emit (which is in fact a property right over the earth's carbon-cycling capacity) to the conscious political control of natural resources.

Political movements do actually challenge this 'fetishism of distribution' by contesting the dominant account and practice of the climate change regime. For instance, across the world, indigenous movements, as well as labor movements, women's movements and so on are uniting under the motto 'leave the oil in the soil'.[44] This movement started first as a reaction to oil company practices that included the appropriation of land, systematic pollutions by gas flaring and spillages and a history of violence towards indigenous and labor movements.[45] However, climate change has become a main focus of these movements and the campaign to stop the exploitation of the remaining oil fields is increasingly concerned with issues of global warming beyond the previous more localized fights.[46] These movements have started discussing and opposing the development of so-called 'carbon offsets' developed under the Kyoto protocol as forms of 'colonialism' or 'imperialism'.[47] More recently, the Cochabamba Summit and its subsequent 'agreement' have highlighted the possibilities of constituting coalitions between indigenous movements and other social movements on an anti-capitalist basis, by contesting the ruthless appropriation of the Earth. One interesting outcome of this conference is the constitution of an alternative ontology of the social and natural world based on the concept of 'Mother Earth' that radically refuses the appropriation of the commons through international law. According to Terisa Turner, 'the World People's Movement [which has arisen out of the Cochabamba Summit] has a clear anti-capitalist analysis reinforced by a sketch of an alternative life-centered, democratic world. Both are informed by practice that is class-based, and indivisibly indigenous, feminist, universalist, socialist, ecologist, and "in the commons, for the commons"'.[48] The point is that these movements challenge the hegemonic account of the climate regime and reveal that, beyond the distribution of greenhouse gases emission rights, struggles are taking place in the sphere of production, over the use-values that are being produced.

THE FETISHISM OF GLOBAL GOVERNANCE

There is a second sense in which the international legal form leads to a fetishist understanding. As mentioned earlier, the international is perceived as the locus of the common interest, standing above narrow national interests. International law is grasped as a form of regulation directed towards the common good. It is invested with the capacity and the will to regulate the 'negative' aspects of 'globalization'.[49] As Saurin writes:

> The portrayal of problems as being global, and their putative solutions, also, as being global, is of key ideological significance . . . The aggregate effect of all states attempting to regulate such public goods is to transfer the worldwide management of environmental public goods into a problem of inter-state environmental regulation.[50]

The immediate effect is to delegitimize the national state as a place of actions. States are perceived as too narrow places and/or too solitary in implementing transformations, and it is argued that those states that would unilaterally implement stronger ecological norms would reduce their economic competitiveness.

Ulrich Brand has identified some recurring features in the hegemonic discourse of global governance that fit well with the fetishism of the governance of climate change. This discourse is characterized, amongst other features, by the idea that 'the problems arising are conceived of as "world problems," a characteristic of which is that they affect all societies and people and therefore everyone must have an interest in dealing with them effectively'.[51] Moreover, these 'global problems' are to be dealt with in a cooperative manner, reflecting the formal equality of sovereignties and leading to win–win situations.[52]

The fetishism of global governance is in line with an understanding of the ecological crisis as a shared burden

> which impl[ies] an equality of responsibility both in causing environmental degradation and in facing the consequences of that global degradation. References to shared 'global' responsibility or to a common fate rely almost wholly on quasi-mystical appeals to some worldwide imagined community which does not and could not have any substantive historical presence.[53]

Mainstream political science has underlined the ways in which such a formal understanding of the climate crisis as a shared burden is 'fuelling injustice'.[54] More deeply however, the formal sovereign equality on which international law is predicated allows for substantial inequalities in access to resources, such as the earth's carbon-cycling capacity. More powerful states have the power to impose criteria for emission allowances that favor them, such as grandfathering.

Actually, some differences between states are acknowledged in international climate change law under the principle of 'common but differentiated responsibilities'. Developing countries, as well as NGOs and others, fought to have this principle enshrined in international law during the Kyoto negotiations. This led to the distinction between Annex 1 and non-Annex 1 countries,[55] between those who do and those who do not have a commitment to greenhouse gases reductions under Kyoto. This is an example in which the international legal form ('common') is overtaken partly by its content ('differentiated') and is the expression of the contradictions that are inherent in the legal form. However, the content disregards the legal form as a relation between countries and does not address deeper structural inequalities which are grounded in the capitalist relations of production: therefore, the content of the legal principle remains undetermined and open to competing interpretations.[56] The failure of the subsequent round of negotiations, especially in Copenhagen in 2009, can be explained precisely because of the struggle between Annex 1 and non-Annex 1 countries. The former insist on the inclusion of the latter in a new protocol that would set reduction targets on all parties.

The fetishism of global governance leads to a form of political inaction. According to Brand:

> The dramatic description of the situation – which prevails in most Global Governance contributions – is answered by very moderate political ideas, which is due to the fact that the constitution of the problems is paid scarcely any attention. The postulated comprehensive claim to profound changes exists alongside the broad acceptance of social relations as they are.[57]

A cosmopolitanism from below is therefore a movement that although aware of the necessary international mediation of its struggle, does not consider such a level as being more in favor of 'general interest of mankind' than the national level, and that therefore seeks action also at other levels of governance, and towards 'non-political' actors such as transnational companies. They thus play counter to the fetishism of global governance in which solutions can come only 'from above'.

For instance, the above-mentioned movements advocate direct action and struggle localized at oil companies and plants. Another interesting example of such a strategy is the 'One Million Climate Jobs Campaign', launched by a social movement with trade union links in the UK. Rather than waiting for international agreements, these union activists ask for public investment in so-called green technologies, public housing, public transport, and so on that would create both 'good jobs' and drastically reduce greenhouse gases emissions in the UK. One of their arguments is indeed a call for direct political action, at a national level, but with a cosmopolitan perspective in mind:

> Of course cuts in the UK on their own will make little difference to global climate change. But if we campaign for a million new jobs, and win them, people all over the world will see what we have done. They will know it is possible. And then they can do the same. And that will save the planet.[58]

THE FETISHISM OF THE STATE

The last fetishism that capitalist relations of production generate, and the one more in line with the cosmopolitan critics voiced in this volume, is certainly the fetishism of the state. By this, I mean, in the context of inter*national* politics, the idea that the states are the pertinent units of allocation of greenhouse gases emissions allowances. By criticizing this idea, again I do not mean that the fetishism of the state is arbitrary or a simple mistake. Rather, I believe that we need to understand that the framing of the discussion on distribution of greenhouse gases around an *inter-states* understanding is itself a necessary part of sovereignty – the formal separation of the economy and of the political – as the expression of capitalist relations of production. The allocation of polluting rights is based on such a form, in which formally equal states have a differential right to pollution. The fact that international negotiations are negotiations between *states* constrains the way in which responsibilities are allocated.[59] The climate regime, as well as most normative discussions regarding the allocation of responsibility, are predicated on the idea that the states are the pertinent units of allocation[60] and that all the citizens of a given state equally are emitting a share of the total emissions and therefore equally are responsible for the production of global warming.

For instance, in an example taken from a popular account of the ecological crisis: 'Each person in the United States is responsible for emitting about six times the world's average emission per person . . . Compared with the poorest countries in the world the responsibility of the average American is staggering.'[61] Obviously, this assertion is never substantiated, in that we are never told that the 'average American' does not exist. What does exist are deeply unequal relations of production, that translate into highly differentiated level of consumptions,[62] but more fundamentally on highly unequal power over the decisions of investment. Constructing the average American against the average inhabitant of a poor country is the best way to avoid discussing relations of production within these states. Further, this allows for a wholly biased understanding of the relations between consumers, citizens and politicians in addressing the problem: 'Democratic politicians, depending on an electorate wedded to increasing car use, more international air travel, and rising consumption, were unwilling to face up to the very difficult, fundamental decisions that would be required to reduce carbon dioxide emissions.'[63]

Therefore, discussing the question of climate change in an inter*national* setting allows for an assessment (and a solution) in terms of national responsibilities: the legal form erases the deep class inequalities within states and therefore also the capitalist relations of production that ultimately drive the global climate crisis.[64] What ultimately are relations between social classes are understood and regulated as relations between states. The orthodox standpoint in Political Science and especially in International Relations reinforces such a view, as 'it accepts as essential and natural the separation of the public and the private realms to which distinct moral and political codes are applicable'.[65] Also, this is why I believe that the quasi-absence of any study on inequalities of domestic production of greenhouse gases between social classes within a given nation is not a mere technical problem but is deeply entrenched in this structural *inter-state* understanding of the problem.[66]

Now, Chakravarty et al. have published a study[67] that is supposed to evaluate per capita emissions of CO_2 at a global level, although their methodology (which approximates individual CO_2 emissions through income) falls well short of the kind of empirical studies that would be needed on this matter. Moreover, it takes as its unit of analysis the 'final' consumer, therefore assuming again that consumption is driving production, re-enacting the above-mentioned fetishism of distribution. This of course leaves out the relations of production and therefore the unequal command over production. To put it in other words, a given capitalist may have a lifestyle that is relatively poor in carbon emissions, but the point is not his own consumption but the fact that, as a member of the capitalist class, he may command over heavy investment decisions that will define in a much more crucial way the total CO_2 emissions amount that what a worker may consume (although this given worker may depend on a heavy amount of CO_2 emissions to live, such as for transportation, and so on). As Saurin puts it:

> the strategic asset of accumulated and concentrated capital allows capitalists to determine the shape, content and direction of future investments irrespective of the needs or conditions of direct labor. Thus whilst direct producers appear to be the immediate agents of environmental change . . . as well as of a corresponding set of social relations of production, their autonomy is structured by principles of appropriation and exchange which they are not at liberty to overturn.[68]

As argued above, this fetishism is not a 'false consciousness'; rather, it is inherent in the social forms that are central to capitalism. In particular, the states-system allows for the kind of legal fetishism discussed above. This has consequences on the kind of social relations that take place within these states. As Raphaël Ramuz writes: 'The concurrence between states reinforces the inner construction of the territorial state by substituting for class antagonisms and capitalist concurrence the construction of a general interest of the (abstract) members of this territorial state within the international concurrence.'[69] This means, for instance, that class struggle is framed in a state perspective, including for movements that point outside the state, for instance towards international solidarity. As Ramuz puts it: 'The existence of competing states allows indeed the building of cross-class coalitions within these states aimed at getting competitive advantages on the world market. This background is necessary to understand policies of "social partnership" or corporatist organizations.'[70] This is obvious in the case of climate change.

A cosmopolitanism from below is therefore the movement towards the creation of solidarities across state borders, that creates links between various social movements, such as trades unions, indigenous movements, environmentalists and women's movements. This form of cosmopolitanism is required to fight the fetishism involved in the inter-state understanding of social relations over the globe. The demonstrations in Copenhagen were made by people from across the world, expressing concerns for humanity as a whole and contesting the nationalist approach embedded in international law. The so-called 'World People's Conference' in Cochabamba was also an attempt at constituting these bonds of solidarity across the borders of the nation-states. These attempts remain fragmented, even contradictory, but they lay the ground for further demonstrations of cosmopolitanism from below in order to fight climate change.

CONCLUSION: COSMOPOLITAN SOLUTIONS

These movements that have appeared over recent years are part of what Ulrich Brand labels 'emancipatory postneoliberal strategies', which may open up 'a way of thinking and acting that go beyond the capitalist mode of societalization'.[71] The class structure in capitalism leads to highly differentiated positions both in the responsibility and in the consequences of the ecological crisis, especially the climate crisis. Cosmopolitan solutions to the climate regime would require taking into account these differences and siding with the struggles of the dominated to overcome domination, and in this course to overcome the fetishisms inherent in the climate regime as structured by capitalist relations of production. This, of course, does not tell us how precisely to devise an alternative organization of the climate regime. However, I believe that political scientists and theorists should pay closer attention to actual social movement and social forces, outside the states, and to the ways in which they understand the climate crisis and the solutions they develop in order to overcome it. This cosmopolitanism from below is what is needed against the fetishism of distribution, of the states and of global governance 'from above', who ultimately are grounded in the reproduction of the very social relations that drive the global ecological crisis.

It seems very dubious to believe that under unchanged capitalist relations of production, an international deal would lead to meaningful reductions in greenhouse gas emissions. Not only is the economy oriented towards restless economic growth, but the states themselves, as instituted by capitalist relations of production, rely on such a growth for their perpetuation. This sets limits to a policy of 'climate capitalism'.[72] Under international law on climate change, the climate regime – although it appears, under a fetishist understanding, as oriented towards the common good – is essentially the expression of these relations of production and their mediations. Cosmopolitanism as a real concern for the fate of every human being, and as the movement towards the realization of this equality, is constrained both by capitalist relations of production, and their mediation through the states. The cosmopolitanism from below expressed in the social movements for climate justice points towards a renewed internationalism, beyond the capitalist relations of production. These movements show us how to make peace with the planet –something utterly absent from the dreadful show we experienced in Copenhagen.

ACKNOWLEDGMENTS

I am grateful to Raphael Ramuz, Dae-oup Chang, Eva Hartmann and Cynthia Kraus for comments on previous drafts of this chapter.

NOTES

1. See the Introduction to this volume, p. 1.
2. I understand 'regime' here in the sense of the set of political regulations constructed around a given 'international problem'.
3. Fuyuki Kurasawa, 'A Cosmopolitanism From Below: Alternative Globalization and the Creation of a Solidarity without Bounds', *European Journal of Sociology/Archives européennes de sociologie* 45, no. 2 (2004): 233–255.
4. Aaron Maltais, 'Global Warming and the Cosmopolitan Political Conception of Justice', *Environmental Politics* 17, no. 4 (2008): 594.
5. Peter Gowan, 'The New Liberal Cosmopolitanism', *IWM Working Paper* 2 (2000): 1–28; Ulrich Brand, 'Order and Regulation: Global Governance as a Hegemonic Discourse of International Politics?', *Review of International Political Economy* 12, no. 1 (2005): 155–176.
6. Maltais, 'Global Warming' (2008): 597. See also Peter Singer, *One World: The Ethics of Globalization*, 2nd ed. (New Haven: Yale University Press, 2004).
7. See Kurasawa, 'A Cosmopolitanism From Below' (2004) and 'Global Justice as Ethico-Political Labour and the Enactment of Critical Cosmopolitanism', *Rethinking Marxism* 21, no. 1 (2009): 83–100.
8. Kurasawa, 'A Cosmopolitanism From Below' (2004), p. 236.
9. Kurasawa, 'Global Justice' (2009), p. 86.
10. *Ibid.*
11. Michael Williams and Geert Reuten, 'The Political Economy of Welfare and Economic Policy', *European Journal of Political Economy* 10 (1994): 265.
12. Karl Marx, *On the Jewish Question* (1843), quoted in Justin Rosenberg, *The Empire of Civil Society: A Critique of the Realist Theory of International Relations* (London: Verso, 1994), p. 69.
13. For the classical exposition of this argument, see Ellen Meiksins Wood, 'The Separation of the Economic and the Political in Capitalism', *New Left Review* 127 (1981): 66–95. See also Philip Corrigan and Derek Sayer, 'How the Law Rules: Variations on Some Themes in Karl Marx', in Bob Fryer, Alan Hunt, Doreen McBarnet and Bert Moorhouse (eds) *Law, State and Society, Explorations in Sociology* (London: Croom Helm, 1981).
14. Ellen Meiksins Wood, 'Logics of Power: A Conversation with David Harvey', *Historical Materialism* 14, no. 4 (2006): 24.
15. Sovereignty is 'the social form of the state in a society in which political power is divided between public and private spheres', Rosenberg, *The Empire of Civil Society*, p. 129.
16. Peter Burnham, 'Marx, International Political Economy and Globalisation', *Capital & Class* 75 (2001): 107.
17. John Holloway, 'Global Capital and the National State', *Capital & Class* 52 (1994): 52.
18. Oliver Nachtwey and Tobias Ten Brink, 'Lost in Transition: The German World-Market Debate in the 1970s', *Historical Materialism* 16, no. 1 (2008): 37–70.
19. Williams and Reuten, 'The Political Economy of Welfare and Economic Policy' (1994), p. 268.
20. China Miéville, *Between Equal Rights: A Marxist Theory of International Law*, vol. 2, Historical materialism book series (Chicago, IL: Haymarket Books, 2005), p. 142.

Social forms are forms of appearance (phenomenal forms) of underlying social relations. As we shall see, the existence of social forms allows for a fetishist understanding of social relations.

21. Jean Louise Cohen, 'Whose Sovereignty? Empire Versus International Law', *Ethics & International Affairs* 18, no. 3 (2004): 15.
22. *Ibid.*, p. 20.
23. *Ibid.*, p. 17.
24. China Miéville, 'The Commodity-Form Theory of International Law', in Susan Marks (ed.) *International Law on the Left: Re-Examining Marxist Legacies* (Cambridge; New York: Cambridge University Press, 2008), p. 120.
25. Miéville, *Between Equal Rights* (2005), p. 293.
26. Tran Hai Hac, *Relire 'Le Capital'. Marx, critique de l'économie politique et objet de la critique de l'économie politique*, vol. 1 (Lausanne: Editions Page deux, 2003), pp. 180–181.
27. Neil Smith, 'Nature as Accumulation Strategy', in *Coming to Terms With Nature*, Leo Panitch and Colin Leys (eds), *Socialist Register* (London, New York, Halifax: Merlin Press, Monthly Review Press, Fernwood Publishing, 2007).
28. Corrigan and Sayer, 'How the Law Rules' (1981), p. 23 (authors' emphasis).
29. *Ibid.*, p. 31 (authors' emphasis).
30. Tran, *Relire 'Le Capital'*, vol. 2, p. 254.
31. Larry Lohmann, 'Marketing and Making Carbon Dumps: Commodification, Calculation and Counterfactuals in Climate Change Mitigation', *Science as Culture* 14, no. 3 (2005): 203–35; and *Carbon Trading: A Critical Conversation on Climate Change, Privatisation and Power*, vol. 48, Development Dialogue (Uppsala: The Dag Hammarskjold Foundation, 2006).
32. Moishe Postone, *Time, Labor, and Social Domination: A Reinterpretation of Marx's Critical Theory* (Cambridge, UK; New York, NY, USA: Cambridge University Press, 1993).
33. In line with the discourse of 'ecological modernization', which Martinez-Alier in his *The Environmentalism of the Poor: A Study of Ecological Conflicts and Valuation* (Northampton, MA: Edward Elgar Publishing, 2002) also calls the 'gospel of eco-efficiency'.
34. And also around the issues of adaptation to and reparation for the consequences of the climate crisis, although I am not discussing these in this chapter.
35. For instance Singer, *One World* (2004).
36. John Bellamy Foster, 'Marx's Theory of Metabolic Rift: Classical Foundations for Environmental Sociology', *The American Journal of Sociology* 105, no. 2 (1999): 366–405; Richard York and Eugene A. Rosa, 'A Rift in Modernity? Assessing the Anthropogenic Sources of Global Climate Change with the Stirpat Model', *International Journal of Sociology and Social Policy* 23, no. 10 (2003): 31–51; Brett Clark and Richard York, 'Carbon Metabolism: Global Capitalism, Climate Change, and the Biospheric Rift', *Theory and Society* 34, no. 4 (2005): 391–428.
37. For an example of such an enquiry see Elmar Altvater, 'The Social and Natural Environment of Fossil Capitalism', in Leo Panitch and Colin Leys (eds) *Coming to Terms With Nature*, Socialist Register (London, etc.: Merlin Press, etc., 2007[Q5]). Also, for the American case see Matthew T. Huber, 'The Use of Gasoline: Value, Oil and the "American Way of Life"', *Antipode* 41, no. 3 (2009): 465–486.
38. Postone, *Time, Labor, and Social Domination* (1993).
39. Altvater, 'The Social and Natural Environment of Fossil Capitalism' (2007).
40. Patrick Bond, 'The State of the Global Carbon Trade Debate', *Capitalism Nature Socialism* 19, no. 4 (2008): 97.
41. *Ibid.*
42. See for instance Maltais, 'Global Warming' (2008), p. 601.
43. Andy Scerri, 'Paradoxes of Increased Individuation and Public Awareness of Environmental Issues', *Environmental Politics* 18, no. 4 (2009): 474.

44. See for instance http://www.oilwatch.org/ or http://www.carbontradewatch.org/ (accessed 30 August 2009).
45. One of the most infamous cases being that of Shell in Nigeria. See Cyril I. Obi, 'Globalisation and Local Resistance: The Case of Ogoni Versus Shell', *New Political Economy* 2, no. 1 (1997): 137–148.
46. Joan Martinez-Alier and Leah Temper, 'Oil and Climate Change: Voices from the South', *Economic & Political Weekly* (2007): 16–19.
47. Heidi Bachram, 'Climate Fraud and Carbon Colonialism: The New Trade in Greenhouse Gases', *Capitalism Nature Socialism* 15, no. 4 (2004): 1–16.
48. Terisa E. Turner, 'From Cochabamba: A World Movement of Social Movements to End Climate Chaos by Ending Capitalism', *Canadian Dimension* 44, no. 5 (2010): 21.
49. Brand, 'Order and Regulation' (2005).
50. Julian Saurin, 'International Relations, Social Ecology and the Globalisation of Environmental Change', in *The Environment and International Relations*, John Vogler and Mark Imber (eds) (London; New York: Routledge, 1996), pp. 92–93.
51. Brand, 'Order and Regulation,' (2005): 160–161.
52. *Ibid.*, p. 161.
53. Saurin, 'International Relations' (1996), p. 82.
54. J. Timmons Roberts and Bradley C. Parks, *A Climate of Injustice: Global Inequality, North–South Politics, and Climate Policy* (Cambridge, Mass: MIT Press, 2007).
55. Participating industrialized countries that are parties to the UNFCCC are referred to as 'Annex 1 countries'. Under Annex B of the Kyoto Protocol, they have a commitment to reduce their carbon emissions. Non-industrialized countries that are parties to the UNFCCC are referred to as 'non-Annex 1 countries'. Under the Kyoto Protocol, 'non-Annex B countries' do not have a binding commitment to the reduction of their carbon emissions. This distinction was based on the principle of 'common but differentiated responsibilities'.
56. Chukwumerije Okereke, 'The Politics of Interstate Climate Negotiations', in Maxwell T. Boykoff (ed.) *The Politics of Climate Change: A Survey* (London; New York: Routledge, 2009), pp. 42–61.
57. Brand, 'Order and Regulation' (2005), p. 168.
58. Jonathan Neale, *One Million Climate Jobs: Solving the Economic and Environmental Crises* (Campaign Against Climate Change, 2010), p. 16.
59. Diana M. Liverman, 'Conventions of Climate Change: Constructions of Danger and the Dispossession of the Atmosphere', *Journal of Historical Geography* 35 (2009): 288.
60. Simon Caney, 'Cosmopolitan Justice, Responsibility, and Global Climate Change', *Leiden Journal of International Law* 18 (2005): 754. Caney discusses four possible units of analysis, namely: (a) individuals, (b) economic corporations, (c) states, (d) international regimes and institutions. Social classes are left out of the discussion.
61. Clive Ponting, *A New Green History of the World: The Environment and the Collapse of Great Civilizations*, revised ed. (New York: Penguin Books, 2007), p. 402.
62. Hervé Kempf, *Comment les riches détruisent la planète* (Paris: Seuil, 2007).
63. Ponting, *New Green History*, p. 406. Note that only the sphere of consumption is alluded to in these examples. The 'consumer' appears as the only instance of power.
64. York and Rosa, 'A Rift in Modernity?' (2003).
65. Julian Saurin, 'Global Environmental Crisis as the "Disaster Triumphant": The Private Capture of Public Goods', *Environmental Politics* 10, no. 4 (2001): 80–81.
66. See Roberts and Parks, *Climate of Injustice* (2007), pp. 284–285 (n. 43).
67. Shoibal Chakravarty, Ananth Chikkatur, Heleen de Colinck, Stephen Pacala, Robert Socolow and Massimo Tavoni, 'Sharing Global CO_2 Emission Reductions Among One Billion High Emitters', *Proceedings of the National Academy of Sciences* (2009): doi=10.1073/pnas.0905232106.
68. Saurin, 'International Relations' (1996), p. 87.
69. Raphaël Ramuz, 'Histoire du capitalisme, logique du capital, ou les deux: Comment analyser l'état comme partie d'un système interétatique?' (paper presented at the 2nd

annual Workshop on Critical Voices in Swiss IR, Lausanne, 2009), p. 19 (author's translation).

70. *Ibid.*
71. Ulrich Brand, 'Environmental Crises and the Ambiguous Postneoliberalising of Nature', *Development Dialogue* 51 (2009): 111.
72. Peter Newell and Matthew Paterson, *Climate Capitalism: Global Warming and the Transformation of the Global Economy* (Cambridge: Cambridge University Press).

6. Sharing the burdens of climate change: environmental justice and qualified cosmopolitanism

Michael W. Howard

INTRODUCTION

Global warming and consequent climate change constitute one of the greatest challenges for our species. Not only are our survival and well-being, and those of future generations, in the balance, the problem is a global one that will require unusual global cooperation and agreement on shared principles. In this chapter I will examine several proposals for principles that should govern the sharing of the burdens of climate change. One idea I will examine is that the polluters should pay for the costs of climate change. I will talk about three versions of this polluter pays principle. But, as we will see, the polluter pays principle by itself is inadequate, because it makes no distinction between polluters who are poor and polluters who are rich. The burdens of climate change will fall heavily on the global poor, in at least two ways: first, the consequences of unmitigated climate change – such as rising sea levels, expanding deserts, more intense storms and disrupted food supplies – will affect areas of the planet inhabited by some of the poorest people, and they lack the resources to adapt. Second, reducing the amount of global warming requires reductions in emissions of greenhouse gases (GHGs, especially carbon dioxide, but also methane and nitrous oxide) just as developing countries need to expand energy use in order to develop, presenting a dilemma between development and climate change mitigation. Justice to the world's poor obliges the wealthier to reduce GHG emissions at a rate that permits development in poor countries, and to assist these countries in meeting necessary GHG reductions of their own. Domestic climate change policy must be shaped not only by global justice but also by a fair sharing of the burden domestically. In both national and global cases, the 'polluter pays' principle should be qualified by an 'ability to pay' principle.

In this chapter I will focus mainly on the global aspect of environmental justice.[1] A distinction is often made between cosmopolitan (or global) justice and international justice. The former focuses on persons as citizens of the world, relegating states to the status of means toward fulfilling the obligations we owe one another as persons, not only within but across borders. International justice is justice between states. Tan makes a further distinction between moral and institutional cosmopolitanism.[2] The former focuses on the duties we owe one another as citizens of the globe, but these duties could be fulfilled by states and international institutions, as well as global institutions. Institutional cosmopolitanism calls for global institutions to displace sovereign states. I am not committed to the latter, but neither am I opposed to it. My focus is on the principles that should govern the sharing of the burdens of climate change, and if I sometimes focus on state structures, it is only because these are more familiar, not because these are the only or the best institutions for realizing global justice.

I shall take for granted the fact, widely acknowledge by climate scientists, of anthropogenic climate change.[3] I shall also accept the standard benchmark, that if we are to avoid catastrophic consequences from climate change, the global temperature must not be allowed to rise more than 2 degrees Celsius above pre-industrial levels, and that such an increase is likely to happen if concentrations of CO_2 in the atmosphere exceed 450 parts per million (ppm).[4] More recent research has suggested that the safe stabilization level should be 350 ppm.[5] Since the concentration is already at 387 ppm, action is needed to cut GHG emissions rapidly and deeply. The likely consequences for human beings of global warming of 2 degrees Celsius or more beyond pre-industrial levels are familiar, but bear repeating. The Intergovernmental Panel on Climate Change (IPCC) has warned that a business-as-usual scenario (tripling of CO_2 to 950 ppm, and a 3–7 degree Celsius temperature rise by 2100) would result in high risks to many unique and threatened systems (including Arctic sea ice, mountain top glaciers, coral reef communities and most endangered and threatened species), large increases in extreme weather events (heat waves, wildfires, droughts, flooding, hurricanes), negative impacts for most regions of the earth, net negative economic impacts (abandonment of or damage to flooded coastal areas or desertified regions, adaptation of infrastructure to new climate conditions, responding to threats from pests and disease) and high risks of large-scale discontinuities (ice sheet collapse, species extinctions and ecosystem collapse, major disruptions in food chains, hundreds of millions of climate refugees).[6]

It will be a challenge to keep the global temperature rise to just 2 degrees.[7] To have a reasonable chance of staying below the 2 degree threshold, global GHG emissions should peak in 2013, and decrease to 80 percent below 1990 levels by 2050.[8] An adequate response to this challenge will be costly, and at this late stage is likely to require measures for adaptation as well as mitigation. How should the burdens be shared? The United Nations Framework Convention on Climate Change (UNFCCC) includes normative principles that I will defend and develop after exploring several conflicting conceptions of fairness in the sharing of climate change burdens.

THE UNITED NATIONS FRAMEWORK CONVENTION ON CLIMATE CHANGE

The United Nations Framework Convention on Climate Change, accepted by 181 countries in Rio in 1992, stipulated that GHGs are to be stabilized at safe levels 'on the basis of equity in accordance with their common but differentiated responsibilities and respective capacities'.[9] This statement includes two principles: first, a principle of responsibility, according to which those who have created the pollution should pay – the polluter pays principle (PPP). Second is a principle of capacity: those who are more able to bear the cost should pay – the ability to pay principle (APP). Both of these principles support the conclusion that 'Developed nations "should take the lead in combating climate change and the adverse effects thereof"'. Developed nations accordingly committed themselves to reductions of emissions to 1990 levels by 2000, a goal which was not fulfilled.[10]

The Framework Convention also endorsed the precautionary principle, 'to avoid the risk of serious and irreversible damage even in the absence of full scientific certainty'. And it recognized a 'right to sustainable development' which justifies initial exemptions from emission reductions for developing countries. The Kyoto Protocol of 1997 set targets for developed countries to limit GHG emissions by 2012 to 5 percent below 1990 levels (8 percent for Europe, 7 percent for the US), and accepted the principle of emissions trading.[11] These targets will not be met. European emissions have continued to rise despite implementation of carbon emissions trading. It remains to be seen whether Europe will achieve its stated goal of 20 percent reductions below 1990 levels by 2020. The US did not ratify Kyoto. The Byrd-Hagel Resolution, which passed the US Senate by a vote of 95–0 shortly before the Kyoto meeting, opposed any exemptions for developing countries.[12] President Bush objected to the agreement because he thought it unfair that 'China and India were exempted from that treaty'.[13] The principle of fairness invoked here, let us call it the 'Future

Polluter Pays Principle', could be expressed as follows: all polluters have an equal obligation not to pollute, regardless of history of past pollution or ability to pay. It would follow that China and India should be under the same emissions targets as developed countries.

History, however, is relevant (at least since 1990), as is ability to pay. To see why some form of historical polluter pays principle should be adopted, consider the question, who has a right to the atmosphere? It is clear that the atmosphere is a commons. So too, once, was most or all of the earth's land. How did the land come to be unequally distributed private property? According to the famous theory of John Locke, when one mixes one's labor with the land, one legitimately takes it out of the commons and makes it proper to oneself. But Locke added an important proviso, that one leave 'enough and as good' for the rest. If this proviso is not met, then the owners owe compensation to those who are excluded, comparable to what they would have had from an equal share of common, unappropriated land.[14] This suggests that in Locke's theory of property there is a right to freedom from extreme poverty, defined as falling below the level of subsistence that one might expect if one had access to one's share of the unimproved commons.

How is this relevant to the atmosphere? Peter Singer compares the atmosphere to a common sink, which for a long time seems to have an infinite capacity, but now is getting clogged. Some have been using it more than others, and if they continue at the current pace, others will be excluded altogether.

> Many of the world's poorest people, whose shares of the atmosphere's capacity have been appropriated by the industrialized nations, are not able to partake in the benefits of this increased productivity in the industrialized nations – they cannot afford to buy its products – and if rising sea levels inundate their farm lands, or cyclones destroy their homes, they will be much worse off than they would otherwise have been.[15]

Singer's analysis supports an obligation for mitigation or adaptation (even on libertarian Lockean grounds), paid for by the polluter, as rectification and compensation for taking more than a fair share. The US has contributed 30.3 percent of cumulative emissions of greenhouse gases, and Europe 27.7 percent, while China, India, and Southeast Asia combined have contributed only 12.2 percent.[16] Singer notes, 'At present rates of emissions . . . contributions of the developing nations to . . . [GHG emissions] will not equal the built-up contributions of the developed nations until about 2030 . . . [Adjusting for population, the per person contributions of developing nations will not equal those of developed nations] for at least another century.' Hence, 'the developed nations broke it' . . . and should fix it.[17]

PROBLEMS WITH HISTORICAL, COLLECTIVIST PPP

There are however problems with a historical, collectivist PPP. The first, recognized by Singer, is excusable ignorance. It is not fair to hold people responsible for emissions that they did not know – and could not have known – were harmful. Excusable ignorance cannot reasonably be pleaded after about 1990, but this still leaves open the question of who should be responsible for emissions prior to this date. Second, earlier generations (distinct from their current descendents) were actual polluters, but they cannot pay.[18] The problem of earlier generations can be addressed if we follow Singer in considering the relevant agents to be collectives – typically countries (although corporations might be included). Then, although currently living American individuals did not emit GHGs in 1920 for example, the US did, and some would say that we as a nation should be responsible for our past collective emissions. Collective agency is problematic, however. As Simon Caney points out, it would be unfair (would violate the principle of equality of opportunity) to allow current residents of a historically high emitting country less than an equal per capita share of current emission rights because of actions of others in their country in the past. In Caney's words:

> It may be true that some people in the past will have had greater opportunities than some currently living people, but that simply cannot be altered: making their descendents have fewer opportunities will not change that. In fact making their descendents pay for the emissions of previous generations will violate equality, because those individuals will have less than their contemporaries in other countries.
>
> So if we take an individualist position, it would be wrong to grant some individuals (those in country A) fewer opportunities than others (those in country B) simply because the people who used to live in country A emitted higher levels of GHGs.[19]

Further, there are affluent polluters in developing countries; why shouldn't they be responsible for some of the burden? And why should poor non-polluters in developed countries be held responsible? These considerations argue for an individualist approach, with 1990 as the reference point. Emissions prior to that date should be treated by an ability to pay principle, which I will discuss shortly.

AN INDIVIDUALIST PPP: EQUAL PER CAPITA SHARES

A more individualist PPP, qualified with respect to excusable ignorance, is the equal per capita shares (EPCS) principle: every person is entitled to an equal per capita share of emission rights, compatible with sustaining an atmosphere for all (now and in the future). What is this per capita share? Peter Singer suggests 1 metric ton of carbon per person per year. This is the amount that would accord with the Kyoto goal of 5 percent below 1990 emission levels in order to stabilize the carbon in the atmosphere at a safe level. (I leave aside whether, in the light of further scientific evidence since the Kyoto protocol, 1 metric ton is still too much.) Actual carbon emissions in the US average 5.4 metric tons per person per year. Japan and Western Europe range between 1.6 and 4.2, with most below 3. Developing countries vary widely, but emit much less per capita than the US, with China at 1.25 and India at 0.32. What follows according to the equal per capita shares principle? India should be allowed to increase its carbon emissions up to a sustainable quota (perhaps with some caveats concerning population growth), China should now reduce its emissions by 20 percent, but the US should reduce its emissions by 80 percent.[20]

This sounds demanding, and may appear politically impossible. These difficulties may in part be reduced through emissions trading: a wealthy country then has the option of reducing emissions by less than its quota, if it pays other countries to take responsibility for the amount.[21] For example, the US could pay India not to emit India's full quota, in order to be able to continue making some emissions above the US quota. In this way a tradable quota becomes an asset for low-emitting countries. Strictly with respect to fairness, Singer observes that the equal per capita shares principle is less demanding on developed countries than the collectivist pure PPP, since it sets aside emissions prior to 1990. On the basis of the EPCS principle:

> for at least a century the developing nations are going to have to accept lower outputs of greenhouse gases than they would have had to, if the industrialized nations had kept to an equal per capita share in the past. So by saying, 'forget about the past, let's start anew,' the pure equal per capita share principle would be more favorable to the developed countries than a historically based principle would be . . . The claim that the [Kyoto] Protocol does not require developing nations to do their share does not stand up to scrutiny. Americans who think that even the Kyoto Protocol requires America to sacrifice more than it should are really demanding that the poor nations of the world commit themselves to a level that gives them, in perpetuity, lower levels of greenhouse gas production per head of population than the rich nations have. How could that principle be justified?[22]

So, we may conclude, the EPCS principle does not go too far. But the question I want to pose is whether the EPCS principle goes far enough. While it takes account of responsibility for post-1990 pollution, it does not address pre-1990 emissions. Equally important, it does not address ability to pay. The importance of this can be illustrated with a thought experiment. Suppose in rich country R everyone has an annual income of US$15,000 or more, and per capita emissions are twice the per capita allotment. Suppose in poor country P no one has an annual income exceeding US$5,000, and half of them live on less than US$2 per day, but, because of their poverty, antiquated technology and slash-and-burn agricultural practices,[23] people in P also emit twice their per capita share of emissions. Imposing equal emissions reduction obligations on R and P means that people in P will have even fewer resources for subsistence needs – they will have lower incomes, and lower productivity because of foregone emissions – so that people in R can enjoy more luxuries. How can people in R justify this to people in P, if it is possible for R to assume more of the costs – greater reductions and/or more of the costs of P's reductions – with the only loss to themselves being a reduction in luxury goods?[24] Of course this example is stylized. No one in R is very poor, and no one in P is very rich. In the real world, there are some very poor people in developed countries and some very rich elites in poor developing countries. The poor in the developed world should have less responsibility for their countries' emissions, and rich polluters in the developing world should not be free of responsibility. Global principles can accommodate this insight even if the institutional framework still assigns responsibility to states, as we shall see.

WHAT THE GLOBAL RICH OWE THE GLOBAL POOR

Hobbesian realists might answer that the people in R are stronger, and the strong do what they can, and the weak suffer what they must (as the Athenians put it to the islanders of Melos). This might be an accurate description of what often happens, but can hardly pass as fair. A fair sharing of the burden would involve R making larger than its equal share of reductions, and taking on some of the cost of P's reductions, so that P still has resources for development. Why is this fair? And just how much of an additional burden should R take on? If I were to flesh out a philosophical account of fairness to support the intuitive idea in the R and P example, I would turn to the influential contractualist theory of justice developed by John Rawls, according to which social justice is what people can expect from each other from the standpoint of an ideal contractual situation which is fair.[25] You can think of this ideal contractual situation as screening out the inequality in the domination of the stronger over the weaker. It is also a way of spelling out what we owe to one another if we regard one another with equal respect and concern. In the ideal contractual situation, no one has more power than another, each is well informed and all put their partiality and biases aside.

To model impartiality, Rawls famously invented the idea of the 'veil of ignorance'. We imagine the parties behind the veil, as they deliberate about what principles of justice should govern their society, being ignorant about what places in society they will occupy. They then have to imagine that they might end up in the worst off social position. This will lead the parties, Rawls argues, to design social institutions and select principles for the allocation of resources, so that the worst off position is as well off as it can feasibly be over time, compared to feasible alternatives. They will, in other words, be guided by a principle – Rawls called it the 'difference principle' – that allows differences of wealth and income and resources only when they are to the advantage of the least advantaged group. So, for example, entrepreneurial activity might yield higher incomes for entrepreneurs, but only to the extent that such activity made the worse off better off than they would be without such incentives in operation. Incomes higher than that would be unjust, because they would enrich the better off from the stock of social wealth, at the expense of the poor.

How does this egalitarian idea apply on the global level? This brings us back to the second question, how much of an extra burden should R take on, of greater emissions reductions than P, and sharing the cost of P's reductions? There is a spectrum of responses, even among liberal egalitarians. But all should agree that ability to pay needs to be taken into account, and those more able have an obligation to take on some of the burdens of climate change, even if not responsible for emissions. Applying this idea on the global level, cosmopolitans point out that country of birth is as much a matter of luck as family wealth or natural talent, and globally just institutions should try to even out the undeserved luck resulting from country of birth in much the same way that national institutions such as public education, or unemployment benefits or universal pensions even out the undeserved luck in society.[26] On the global level, the undeserved inequalities are massive. I do not have time to catalogue them, but suffice it to note that 30,000 children die every day from preventable causes, and half of the world's population struggles to survive on US$2 a day or less.[27] If global economic inequalities were held to the same standard that egalitarians think should hold within societies, then the problem of sharing the burdens of climate change would be subsumed under a comprehensive principle for sharing all of the benefits and burdens of global society. This would involve much more substantial transfers of wealth than those discussed thus far with respect to GHG emissions. But making the wealthy countries shoulder most of the burden of climate change could be seen as a step on the way toward global justice.

As appealing as cosmopolitan egalitarianism is to some of us – particularly if we are already egalitarians with respect to national justice – there is considerable controversy about this idea even among egalitarians. I can only give a hint of what this controversy is about here. The core objection is that people who share a 'basic structure', a state, political institutions, shared deliberations, a common culture, or who impose coercive laws on one another owe to each other a higher standard of distributive justice than they owe to those who do not share the same country, in something like the way that we owe more to our children or business partners than we owe to unrelated strangers.[28] Non-cosmopolitans often concede that we owe some human rights minimum to strangers: we have duties not to harm them, or to come to their assistance in times of extreme need, but we do not owe them a fair share of the benefits of life in common. Cosmopolitans retort that in the globalized economy, we have all become citizens of a world shared in common. A basic structure has emerged on the global level, through the increasing dominance of such institutions as the IMF and WTO.[29] Non-cosmopolitans retort in turn that we are not quite there yet.

This is not the place to sort out that debate. For purposes of addressing climate change, I am proposing something much more modest than unqualified cosmopolitan egalitarianism, but perhaps more robust than the human rights minimum that non-cosmopolitans favor, and probably more than the realists would concede. My position is cosmopolitan in taking persons rather than states to be the relevant objects of moral concern. But with Darrel Moellendorf I take the Rawlsian idea that principles must track institutions. Where there are no associational ties, there are humanitarian obligations, but no duties of justice.[30] Moellendorf argues, contrary to Rawls, that there is enough of a basic structure on the global level to ground norms of justice. What I am proposing, however, is not a comprehensive global principle, such as global equality of opportunity, or a global difference principle. Rather, norms of justice are restricted to the goods and practices that are shared on the global level. In this approach, egalitarian principles (going beyond humanitarian assistance) come into play when institutions come into existence that are difficult to escape, affect others and involve some shared governance, even if not full sovereignty, but the principles govern only the goods and harms at issue in the institutions concerned (for example, the benefits of trade, or the costs of global warming).

I can illustrate this restricted cosmopolitan egalitarianism with the example of fair trade.[31] Fair trade is more than free trade. Free trade involves terms of trade that enable parties to trade to their mutual advantage. But mutual advantage is compatible with some parties, because of their greater power, getting more of the advantages and others getting less: free trade need not be fair. One way of making sense of demands for fair trade is to see them as invoking consideration for the worse off, not in the full sense that cosmopolitans demand, but simply with respect to the gains of trade. The terms of trade, the rules of the game, should be structured so as to give priority to the interests of the worse off, and in practice this means making allowance for development needs. This might, for example, allow exemptions from rules governing intellectual property, or permission to give protection to infant industries. But obligations to the poor across borders are not limitless; they are focused on the distribution of the gains of trade.

Typically fairness with respect to trade is concerned with sharing of benefits (although there are also some harms to deal with). Fairness with respect to climate change is primarily about sharing burdens – those of transition to a non-carbon economy – namely, foregoing GHG emissions and investing in new technologies. In a fair contract situation parties will consider what it would be like to be in the situation of the others, particularly the worse off, and would favor arrangements that shift some of the burden to those more able to pay, so that the worse off will be less disadvantaged by the shared burden. This falls short of full cosmopolitan egalitarianism, because the wealthy are not required to bring the world's poor up to the highest feasible standard of living. But the wealthy should burden the poor as little as possible in the shared practice of mitigating and adapting to climate change, until the poor cross a reasonable threshold of development. How far this position is akin to the human rights minimum of non-cosmopolitan liberal nationalists, or goes beyond that position in the direction of cosmopolitan egalitarianism, depends on how this threshold is defined. Both cosmopolitans and liberal nationalists should conclude that an APP must complement the PPP, although they may differ concerning how much weight to give to the APP.

The Greenhouse Development Rights (GDR) framework I discuss in the next section is consistent with the qualified cosmopolitan position I have described, but might also be defended on the basis of a human rights minimum, or as a step toward full global egalitarianism.

GREENHOUSE DEVELOPMENT RIGHTS

The Greenhouse Development Rights framework (developed by Paul Baer et al. of the environmental institute EcoEquity) illustrates one way of combining a PPP with an APP.[32] This hybrid account accords with and fleshes out the norms of equity, sustainable development, responsibility and capacity articulated in the Framework Convention. Recall that the goal is to reduce emissions fast enough and far enough to keep the temperature rise below 2 degrees Celsius. What will this require of developed and developing countries? To allow any scope for developing countries to develop, which at the present time cannot happen without some further increased use of fossil fuels, the emissions in the developed countries must peak sooner than in the developing countries, and decline quickly. This will allow room under the 2 degree limit for developing countries to develop with less demanding emissions restrictions, peaking later and ultimately converging with developed countries at lower emission levels. Room to develop is imperative, given the poverty in the developing world. To repeat our earlier intuition, there is no compelling reason why the world's poor should be denied the right to develop, so that the world's affluent can enjoy more luxury.

The GDRs framework incorporates both an APP (capacity) and a PPP (responsibility), each qualified with reference to development thresholds. Capacity consists of resources beyond what is necessary for basic needs, here defined as income above US$20 per day purchasing power parity. A polluter pays principle applies to post-1990 emissions, excluding emissions of those individuals below the development threshold. Some human rights minimalists might think that this threshold is too high, whereas some cosmopolitans, particularly if not restricting their egalitarianism to the burdens of climate change, might argue that it is not high enough. Only incomes above the threshold, wherever it is set, would count in determining ability to pay. India has very little, China more and the US most of its income above the threshold. Countries would be considered responsible for reducing only emissions beyond 'subsistence emissions', that is emissions by those who fall below the subsistence threshold. Emissions of these people are typically much less than emissions of wealthier people. Most of India's emissions would be exempt, quite a few of China's, but few of those of the US. The Responsibility-Capacity Index (RCI) takes account of emissions and incomes above the thresholds. Note that wealthier countries begin with a higher initial RCI – and thus a greater share of the responsibility for emissions reductions – which declines as the RCI for developing countries rises. Thus, for example, the US and the EU will initially have higher targets, in total and proportionately, than India or China. The US, with 4.5 percent of the world's population, and a GDP per capita of US$45,640 (purchasing power parity), has 25.7 percent of global capacity, 36.4 percent of global responsibility and an RCI of 33.1. The respective numbers for the EU 27 are 7.3, US$30,472, 28.8, 22.6 and RCI 25.7. For

India the respective numbers are 17.2, US$2,818, 0.86, 0.30 and RCI 0.48, and for China they are 19.7, US$5,899, 5.8, 5.2 and RCI 5.5.

But later the obligations of developing countries will increase, as more of their people rise above the development threshold. The RCI for the US declines to 29.1 in 2020, and 25.5 in 2030. The EU RCI also declines, but the RCI for China rises to 10.4 and then 15.2, and India's RCI rises to 1.18 and then 2.34. In this scenario the US should aim for a 6.7 percent decline of emissions annually and 90 percent reductions below 1990 levels by 2050, 'greater than those mandated by even the strictest of the bills in play in the US'.[33] For comparison, the Waxman-Markey climate change bill that passed the House of Representatives in June 2009 envisions cuts of 83 percent by 2050, but aims only for 'modest reductions of 17% from 2005 levels by 2020' equivalent to 1 percent below 1990 levels. The bill also gives away, rather than auctions, 85 percent of the carbon permits, and weakens regulation on farm practices. More modest early reductions entail steeper cuts later in excess of 6.7 percent per year to reach the 2050 target, resting on the optimistic 'assumption that technology to mitigate emissions will improve'.[34] One implication of the GDRs analysis is that the US, if required to reach GDRs targets entirely through reduction of US emissions, could not succeed, as it would need to reduce emissions by more than 100 percent by 2025! It is therefore critical that the scheme includes technology transfer and support for mitigation in developing countries, and probably emissions trading.

China will ultimately face necessary emissions reductions of its own. There is evidence of China's willingness to embrace emissions cuts despite its official bargaining position to the contrary, motivated by concern for the consequences for China of unmitigated climate change. 'One government study has warned that higher temperatures and volatile rainfall could cause production of rice, wheat and corn – the staples of the Chinese diet – to fall 37% by 2040 unless effective adaptation measures are taken.'[35] But under the GDRs framework China would be helped in reaching its targets with support from developed countries.

COSTS

There are widely ranging estimates of the costs for adaptation and mitigation, based on different emissions targets, and different assumptions about the costs of new technologies and the positive benefits of renewable energy. The IPCC thinks the costs of stabilizing in the 445–535 ppm CO_2e (CO_2-equivalent) range would be less than 3 percent of GWP in 2030.[36] In an upward revision of his estimate in The Stern Review, a widely cited economic analysis, Nicholas Stern estimates that to stabilize 'below 500 ppm would . . . cost around 2% of GDP'.[37] European estimates are closer to 0.25 percent of 2020 GWP.[38] The GDRs emergency pathway aims for a lower stabilization level (400ppm CO_2e by 2100), so costs will be somewhat higher. Joseph Romm offers a much more optimistic assessment, based in part on estimates of large savings to be had from energy conservation.[39] For purposes of illustration, suppose costs in 2020 of 1 percent of GWP; then the national obligations of the US would be US$275 billion. This is 1.51 percent of GDP – higher than the global percentage because of our greater capacity and responsibility. This amounts to a per capita cost of US$841 for everyone in the US above the development threshold. Distributed progressively, the costs would be higher for people making US$60,000 per year, but lower for those in lower brackets. And of course, if the total cost is closer to 2 percent of GWP, the costs will be higher, but if the costs are less than 1 percent the costs in each bracket will be lower.[40]

Such costs are substantial, and politically very challenging, but clearly affordable without reducing our standard of living to the level of developing countries, or even giving up much of what we consider important. While my case has been presented in terms of what is required by global justice, it is worth noting that a similar outcome could be defended from a realist perspective: that reducing global GHG emissions will require international cooperation between rich and poor countries, and that getting compliance will require an approximation to some such standard of fairness. That indeed is the claim of the proponents of the GDRs framework.[41]

OTHER APPROACHES

Simon Caney offers a cosmopolitan approach similar to the GDRs framework, designed to solve problems arising for Singer's PPP. Recall that Singer doesn't address emissions from earlier generations, or the problem of non-compliance. Caney addresses these problems by combining a PPP with an APP in the following prescription of duties:

D1: All are under a duty not to emit greenhouse gases in excess of their quota.

D2: Those who exceed their quota (and/or have exceeded it since 1990) have a duty to compensate others (through mitigation or adaptation) (a revised version of the 'polluter pays' principle).

D3: Regarding greenhouse gas emissions resulting from earlier generations, excusable ignorance, and non-compliance: 'the most advantaged have a duty either to reduce their greenhouse gas emissions in proportion to the harm resulting from [earlier generations, etc.] (mitigation) or to address the ill-effects of climate change resulting from [earlier generations, etc.] (adaptation) (an ability to pay principle).'

D4: 'The most advantaged have a duty to construct institutions to discourage non-compliance (an ability to pay principle).'[42]

In contrast with a pure PPP, this 'hybrid account' prioritizes the interests of the global poor. For a country like China, which has high emissions but is poor, the PPP would impose the same duty to reduce emissions as that imposed on wealthy countries. (This is what Bush and the Senate wanted.) The hybrid account will impose some obligations on poor countries like China, but less demanding than a pure PPP. For a country with relatively low emissions, but rich (Sweden), the PPP would impose fewer obligations, but the hybrid account would impose obligations in accordance with the country's ability to pay.

Whether Caney's or the GDRs' framework is more advantageous to the least advantaged depends on the relative quantities of emissions before and after 1990, and how much different groups emit, a comparison I shall not undertake here. Caney subsumes all pre-1990 emissions under the APP principle (D3), but post-1990 emissions responsibilities are allocated according to the equal shares per capita PPP. The GDRs framework exempts post-1990 emissions that are under the development threshold, and so builds in a further development priority qualifying PPP. But GHGs already in the atmosphere from pre-1990 emissions appear to be treated as a background condition shared equally by all. One might object that the inhabitants of historically big polluting countries benefit from past pollution, and so should have some responsibility for it.

One might try to add a beneficiary pays principle (BPP) to meet this objection. There are some philosophical problems with this principle. As Caney puts it, 'We cannot say to people "You ought to bear the burdens of climate change because without industrialization [in the past] you would be much worse off than you currently are." We cannot because without industrialization the "you" to which the previous sentence refers would not exist.'[43] And a forward-looking priority for the worse off may do the work expected of such a principle without the need to assign responsibility to current individuals for past pollution by other individuals. The GDRs framework could be brought into alignment with Caney's hybrid model (if they diverge) by adjusting the weights given to capacity and responsibility in the Responsibility Capacity Index. Baer et al. give these equal weight.

Although it has not been the focus of this chapter, I mention in passing a third type of cosmopolitan approach that combines APP and PPP, assigning responsibility not to countries, but directly to rich and high-emitting individuals.[44] While the proposal I favor in this chapter, the Greenhouse Development Rights Framework, is presented as an international rather than a global institutional scheme, the guiding principle is cosmopolitan: that rich polluting individuals should bear responsibility for global warming, not wealthy countries. That is why individuals below the development threshold, whether in developed or developing countries, are not considered to have the ability to pay, and are not responsible for the cost of their emissions. If a global institutional scheme can be devised that is more effective and politically feasible than the international GDRs framework, all the better.

Vanderheiden rejects an ability to pay principle altogether, in favor of a polluter pays principle, grounded in an account of responsibility that encompasses corrective as well as distributive justice (see Chapter 2). His approach may prove particularly fruitful as attention shifts from mitigation to adaptation, and to compensation for harm resulting from mitigation and adaptation failures. And a negative duty not to harm, which he articulates, may make a stronger claim on us than a positive duty to share in benefits and burdens. The practical results for mitigation of his approach and that defended here are quite similar, although the reasons for them are somewhat different. This chapter focuses mostly on mitigation, and pre-1990 emissions are dealt with under the rubric of distributive justice, as bad luck that we all must share. From a contractualist, egalitarian point of view, such burden sharing must give priority to the worst off, and take account of ability to pay. Vanderheiden assigns responsibility to individuals on the basis of their post-1990 emissions, as do I. But he appears to make these same individuals responsible for the pre-1990 emissions of the collectives (states) of which they are members, which invites the objections I raised earlier about collective responsibility and excusable ignorance. Perhaps a corrective justice analysis can provide answers to these objections better than a distributive justice analysis, but those are objections such a position must address.

In exempting 'survival emissions', or only assigning fault to 'non-impoverished persons', he is acknowledging the same priority for the worse off that motivates a hybrid account such as the GDRs framework. Vanderheiden's approach incorporates 'ability to pay' by recognizing 'survival emissions' as justified as a human rights minimum, trumping equal per capita shares. Elsewhere, he recognizes that a case can be made for not only survival emissions but also development rights that would allow poor people some luxury emissions beyond the emissions necessary for survival, with the caveat that the rich would need to reduce emissions proportionately to allow space for development within an overall regime that kept total GHG emissions within safe limits. The example he discusses is the

GDRs framework defended in this chapter.[45] How far his recommendations differ from the GDRs framework depends on how generously 'survival emissions' are defined, or how far beyond 'survival emissions' are the emissions that would be justified by development rights. The moral hazard problem that Vanderheiden raises for Caney's hybrid approach, 'letting mostly affluent polluters off the hook by transferring some of their liability to those affluent persons that have taken pains to reduce their carbon footprints', could be addressed at the domestic policy level, for example with a carbon cap-and-dividend scheme or other measures such as I discuss in the next section.

CONCLUSION: SHARING THE DOMESTIC BURDEN

I want to conclude with a few remarks about sharing the costs of climate change domestically. While the focus of my talk has been global justice, the costs of global justice are likely to be distributed to countries, and people within each country must share these costs among themselves. Moreover, climate change policies have to be developed within each country, even if driven by international treaties. Here, the philosophical case is somewhat less difficult, even if the political task is arduous. On a global level, there is controversy, both politically and philosophically, about what the global rich owe the global poor. Is it a decent minimum, or something more? But in a national context, there is much wider agreement with the idea, shared by widely varying theorists of social justice, that we owe each other more than a decent minimum; we strive to provide equality of opportunity. We owe each other duties of distributive justice because we share a basic structure – a legal and economic system that determines many aspects of each individual's life. Such duties are implicit in the practice of universal retirement pensions, guaranteed health benefits (at least for the poor and elderly), universal education and other benefits of modern states that are not extended to non-residents. In such a context, it should be evident that costs of climate change will not justly be set only by a PPP, but should take account of ability to pay.

Policies to reduce GHG emissions include carbon taxes, such as have been introduced recently in British Columbia and, more prominently, cap and trade schemes, which have been introduced in Europe, and formerly in the US, with some success, in dealing with the ozone hole.[46] The idea in all of these schemes is to make use of market mechanisms to encourage conservation and shift demand away from fossil fuels to non-GHG emitting sources of energy, by raising the cost of fossil fuels. Carbon taxes do this most directly and transparently, but cap and trade is featured in most legislative proposals. President Obama's 2008 campaign proposal was for a carbon permit auction that would have put a price on all carbon emissions, and he also proposed distributing a large part of the revenue in the form of a dividend to citizens.[47] It makes sense to set aside some of the revenue for investment in renewable energy, energy efficiency and the like. But the dividend is important to insure that the cost of the transition does not fall regressively on the poor, who spend a higher percentage of household income on necessary energy costs. It would also be good politics to gather support for climate change policies if everyone, and particularly low- and middle-income households, received a monthly check as a rebate for the increases in energy costs due to carbon permit fees. In the spirit of GDRs, the dividend would be one way of exempting everyone below the threshold from the burden of increased energy costs.

At the same time, since prices of fossil fuel use will rise for everyone, rich and poor alike will find it in their interests to conserve energy and reduce emissions. Unfortunately, the climate change legislation that passed the House of Representatives in June 2009 gives away, rather than auctions, 85 percent of the carbon permits. Defenders say that this does not matter for the goal of emissions reduction, since the important thing is to set a cap. But it does matter for environmental justice, because the revenue from the cap and trade scheme is thereby reduced by 85 percent, and the amount available from the remaining 15 percent that might be used for a dividend is much smaller. This amounts to a regressive give-away to energy companies at the expense of the domestic low and middle-income population. It could also threaten the political viability of a national GHG emissions reduction program, and indirectly the political viability of a strong international climate change protocol emerging after the Copenhagen meeting in December 2009. Such an outcome would be a loss for all parties. Recognizing these shortcomings, it is nonetheless important to acknowledge that the bill is a first step for the US making real commitments toward reduction of GHGs, and that the bill does have rebates to help low income people with rising energy costs.[48]

Meeting the challenge of climate change will require large reductions in GHG emissions in the coming decades, by all countries. But fair sharing of this burden will require initial exemptions for the poor below the development threshold, larger reductions from the rich, and help for the poor in meeting their reduction targets. Linking the problem of climate change to the problem of development is more likely to succeed than treating climate change in abstraction from development. GDRs afford us a reasonable framework for thinking about environmental justice, and one that accords with the UN Framework Convention. Perhaps this accord, together with the urgency of the threat of climate change, allows us some hope that some such framework will be adopted.

NOTES

1. For an earlier paper in which I give more attention to national policy, see 'An Open Letter to President Obama on Environmental Justice', 2009, available at http://www.usbig.net/papers.php (accessed 15 July 2010).
2. Kok-Chor Tan, *Justice without Borders* (Cambridge: Cambridge University Press, 2004), p. 10.
3. On the scientific consensus see Naomi Oreskes, 'The Scientific Consensus on Climate Change', *Science* 306 (2004): 1686; P.T. Doran and M. Kendall Zimmerman, 'Direct Examination of the Scientific Consensus on Climate Change', *EOS* 90 (2009): 22.
4. Intergovernmental Panel on Climate Change, *Climate Change 2007: The Physical Science Basis*, Contribution of Working Group I to the Fourth Assessment Report of the Intergovernmental Panel on Climate Change (Cambridge: Cambridge University Press, 2007); Stephen Schneider, 'The worst-case scenario', *Nature* 458 (April 2009): 1104–1105; Pew Center on Global Climate Change, 'Key Scientific Developments Since the IPCC Fourth Assessment Report', Science Brief 2, June 2009; Union of Concerned Scientists, 'Findings of the IPCC Fourth Assessment Report: Climate Change Science', last revised 16 February 2007, available at www.ucsusa.org/global_warming/science_and_impacts/science (accessed 30 June 2009).
5. James Hansen, Makiko Sato, Pushker Kharecha, David Beerling, Valerie Masson-Delmotte, Mark Pagani, Maureen Raymo, Dana L. Royer and James C. Zachos 'Target Atmospheric CO^2: Where Should Humanity Aim?' *The Open Atmospheric Journal* 2 (2008): 217–231.
6. Union of Concerned Scientists (2007).
7. A useful overview of how emissions can be leveled and reduced during the next 50 years to avoid a doubling of CO_2, through 'stabilization wedges' – a menu of strategies for reducing CO_2 in the atmosphere with available technologies – is available from the Carbon Mitigation Initiative of the Princeton Environmental Institute, available at http://www.princeton.edu/%7Ecmi/resources/stabwedge.htm (accessed 30 June 2009).
8. Paul Baer, T. Athanasiou, S. Kartha and E. Kemp-Benedict, 'The Greenhouse Development Rights Framework', 2nd ed., November 2008, available at http://gdrights.org/2009/02/16/second-edition-of-the-greenhouse-development-rights/ (accessed 30 June 2009).
9. Quoted in Peter Singer, 'One Atmosphere', in *The Global Justice Reader*, Thom Brooks (ed.) (Oxford: Blackwell Publishing, 2008): 671. First published in Peter Singer, *One World: The Ethics of Globalization*, 2nd ed. (New Haven: Yale University Press, 2002), pp. 14–50, 205–208.
10. According to the Energy Information Agency, US CO_2 emissions rose 20 percent from 4.99 billion metric tons in 1990 to 5.98 metric tons in 2005. European emissions increased at a lower rate, from 4.1 billion to 4.38 billion metric tons over the same period. China's emissions more than doubled, from 2.24 to 5.32 billion metric tons. Available at http://www.eia.doe.gov/oiaf/1605/ggrpt/index.html (accessed 30 June 2009).
11. Singer (2008), p. 671.
12. Baer et al. (2008), p. 111.
13. Quoted in Singer (2008), p. 673.
14. John Locke, *The Second Treatise of Government* (Indianapolis: Bobbs-Merrill, 1952 [1690]), pp. 16–30.
15. Singer (2008), p. 675.

16. Al Gore, *An Inconvenient Truth* (Emmaus, Penn.: Rodale, 2006).
17. Singer (2008), p. 677.
18. Simon Caney, 'Cosmopolitan Justice, Responsibility, and Global Climate Change', in *The Global Justice Reader*, Thom Brooks (ed.) (Oxford: Blackwell Publishing, 2008), p. 703, first published in *Leiden Journal of International Law* 18 (2005). Caney mentions a third problem, the problem of non-compliance: 'those who do not comply with their duty not to emit excessive amounts of GHGs (will not pay)'. I will not discuss this problem further, except to note that among the duties Caney lists below, D4 is designed to address this problem.
19. Caney (2008), p. 701. See also Singer (2008), p. 682.
20. Singer (2008), p. 678. I have updated Singer's figures with data for CO_2 emissions for 2006 from The Union of Concerned Scientists, 'Each Country's Share of CO_2 Emissions', available at: http://www.ucsusa.org/global_warming/science_and_impacts/science/each-countrys-share-of-co2.html (accessed 30 January 2010); 'Carbon Emissions per person by country', *The Guardian* (2 September 2009), available at: http://www.guardian.co.uk/environment/datablog/2009/sep/02/carbon-emissions-per-person-capita (accessed 28 February 2010); US Energy Information Administration, 'International Energy Outlook 2009', available at: http://www.eia.doe.gov/oiaf/ieo/emissions.html (accessed 30 January 2010); The World Bank, 'CO_2 Emissions: Metric Tons Per Capita', available at: http://datafinder.worldbank.org/co2-emissions (accessed 30 January 2010).

    The ratio of CO_2 to carbon is 3.67 according to Dick Hill, 'Carbon Confusion', *Bangor Daily News* (11–12 July 2009).
21. Singer (2008), p. 683.
22. *Ibid.*, p. 682.
23. Gore (227) reports that 'Much of the forest destruction comes from burning. Almost 30% of the CO_2 released into the atmosphere each year is a result of burning of brushland for subsistence agriculture and wood fires used for cooking.'
24. See Singer (2008), pp. 679–680.
25. John Rawls, *A Theory of Justice* (Cambridge: Harvard University Press, 1971).
26. Philippe Van Parijs, 'Global Distributive Justice', in *Blackwell's Companion to Political Philosophy*, R. Goodin, P. Pettit and T. Pogge (eds) (Oxford: Blackwell, 2006).
27. Thomas W. Pogge, *World Poverty and Human Rights: Cosmopolitan Responsibilities and Reforms* (Cambridge: Polity Press, 2002).
28. Thomas Nagel, 'The Problem of Global Justice', *Philosophy & Public Affairs* 33, no. 2 (2005): 113–147; David Miller, 'Against Global Egalitarianism', in *Current Debates in Global Justice*, G. Brock and D. Moellendorf (eds) (Dordrecht, The Netherlands: Springer, 2005), pp. 55–79; John Rawls, *The Law of Peoples* (Cambridge, Mass.: Harvard University Press, 1999).
29. Charles Beitz, *Political Theory and International Relations* (Princeton: Princeton University Press, 1979); Darrel Moellendorf, *Cosmopolitan Justice* (Boulder, Colorado: Westview Press, 2002).
30. *Ibid.*
31. Aaron James, 'Distributive Justice without Sovereign Rule: The Case of Trade', *Social Theory and Practice* 31, no. 4 (2005); available at https://webfiles.uci.edu/ajjames/DistributiveJusticewithoutSovereignRule.pdf; Michael W. Howard, 'Cosmopolitanism, Trade, and Global (or Regional) Transfers', presented to the Basic Income Earth Network, Dublin, Ireland (June 2008), available at http://www.basicincome.org/bien/pdf/dublin08/5dihowardcosmopolitanism.doc (accessed 15 July 2010).
32. Baer et al. (2008). Readers are encouraged to view the graphs that accompany this paper at the previously cited website.
33. Baer et al. (2008), p. 77.
34. 'In Need of a Clean', *The Economist* (27 June 2009), available at http://www.economist.com/world/unitedstates/displaystory.cfm?story_id=13933204 (accessed 30 June 2009); Mark Hertsgaard, 'Shades of Green', *The Nation* 289 (20/27 July 2009): 4–6.

35. *Ibid.*
36. Baer et al. (2008), p. 109, n. 47.
37. Nicholas Stern, *The Economics of Climate Change: The Stern Review* (Cambridge: Cambridge University Press, 2007); Juliette Jowit and Patrick Wintour, 'Cost of tackling global climate change has doubled, warns Stern', *The Guardian* (26 June 2008), available at http://www.guardian.co.uk/environment/2008/jun/26/climatechange.scienceofclimatechange (accessed 13 February 2011).
38. Baer et al. (2008), p. 109, n. 47.
39. See Joseph Romm, Climate Progress (blog), available at http://climateprogress.org/2009/03/30/global-warming-economics-low-cost-high-benefit/ (accessed 30 June 2009).
40. Baer et al. (2008), p. 61.
41. *Ibid.*, pp. 91–92.
42. Caney (2008), pp. 704–705.
43. *Ibid.*, pp. 695–696.
44. Shoibal Chakravarty, Ananth Chikkatur, Heleen de Coninck, Stephen Pacala, Robert Socolow and Massimo Tavoni, 'Sharing Global CO_2 Emission Reductions among One Billion High Emitters', *Proceedings of the National Academy of Sciences* 106 (6 July 2009). For a summary, see Moises Velasquez-Manoff, 'One Way to Decide How Nations Reduce their Carbon Footprint', *Christian Science Monitor* (8 July 2009). See also Paul G. Harris, *World Ethics and Climate Change: From International to Global Justice* (Edinburgh: Edinburgh University Press, 2010).
45. Steve Vanderheiden, *Atmospheric Justice: A Political Theory of Climate Change* (New York: Oxford University Press, 2008), pp. 247–252.
46. Nicola Jones, 'North America's First Carbon Tax Faces Judgement', *Nature News* (5 May 2009), doi:10.1038/news.2009.445 News.
47. A climate dividend has also been supported by Robert Reich, 'Why Revenues from Cap-and-Trade Should Be Returned to Us As Dividends', (4 June 2008), available at http://robertreich.blogspot.com (accessed 15 July 2010). For a lengthier analysis, see Robert Greenstein, Sharon Parrot and Arloc Sherman, 'Designing Climate-Change Legislation that Shields Low-Income Households from Increased Poverty and Hardship', Center on Budget and Policy Priorities (9 May 2008), available at www.cbpp.org (accessed 30 June 2009).
48. Pew Center on Global Climate Change, 'Myths about the Waxman-Markey Clean Energy Bill', Climate Policy Memo 2 (June 2009). See Joseph Romm, 'Why Warren Buffett is Wrong about Cap and Trade', Climate Progress (14 July 2009), available at http://climateprogress.org/2009/07/14/why-warren-buffett-is-wrong-about-cap-and-trade/#more-9078 (accessed 15 July 2009).

7. Cosmopolitanism and hegemony: the United States and climate change

Robert Paehlke

On the surface it would seem that hegemonic power in the hands of a nation would encourage a highly internationalist, even if not necessarily a cosmopolitan, outlook within its population. What could be worldlier than hegemony? Today's hegemonic power, the United States, might also be assumed to have an interest in maintaining global stability since stability is crucial to maintaining its economic and strategic dominance. Yet, until recently, at the national political level the United States hardly even acknowledged one of the greatest threats to stability – climate change. Even today, after the 2010 elections, there is no assurance that the United States will act effectively to protect climate stability and thereby global political and economic stability. Indeed, many of its political leaders and citizens see no reason to act on this issue. To understand the politics of climate change we need to understand why this seeming contradiction continues to exist and why a nation with so much to lose remains reluctant to act. At the level of the 'security state' the United States is well aware of the threats to global stability posed by climate change. The US Department of Defense, for example, has paid close attention to the strategic implications of climate change. Some within the security policy community have concluded that: 'If the United States does not lead the world in reducing fossil-fuel consumption and thus emissions of global warming gases . . . a series of global environmental, social, political and possibly military crises loom that the nation will urgently have to address.'[1]

Security analyst Gwynne Dyer has concluded that political crises related to food production, drought and/or flooding are likely future outcomes of climate change.[2] In 2009, the US Central Intelligence Agency began to provide access to satellite data and photos of Arctic ice melt to climate scientists seeking thereby to advance understanding of the impacts of climate change on ice melt and sea levels.[3] Somehow this concern at the level of the national security apparatus has not significantly influenced the thinking of the self-described national security (Republican) party. At the same time many leading US corporations also favor action on climate change, as do many US states and most large municipalities. Yet President Obama, who campaigned on climate change action and has appointed officials who are strong advocates of climate change action, would not commit in Copenhagen to decisive action especially in the near term.

President Obama chose not to put forward any more than what, at that time, the House of Representatives had tentatively approved and the United States Senate might plausibly accept on a bipartisan basis. The level of US greenhouse gas (GHG) reductions that he put forward is seen as insufficient by most climate scientists. More than that, the January 2010 election of a Republican Senator in normally Democratic Massachusetts and the 2010 general elections made even this modest proposal seem highly problematic. Obama's caution, it turns out, was an apt assessment of many American's instinctive caution regarding even a modest climate change agenda. So deep is America's ideological and structural resistance to a cosmopolitan perspective on this and other issues that it could ultimately undermine America's capacity for global leadership. Climate change is thus a crucial challenge to the Obama administration as well as to the world as a whole. Without leadership on the part of the United States it will be very difficult to convince many nations outside of Europe to take significant climate change action. As to the need for US leadership one needs only to look at the general failure of the Kyoto Accord absent active US participation – global GHG emissions have continued to rise since the agreement was signed.

Regarding Obama's caution in Copenhagen one need only look at the failure of the US to remain on-side after the Clinton administration had agreed in Kyoto to participate. What many forget is that several months *prior* to the acceptance of the Kyoto agreement by Vice President Gore and President Clinton, the United States Senate voted unanimously (95–0) for a resolution that expressed caution regarding the principal of differentiated responsibilities, a principal that placed a stronger burden of immediate action on those nations that had historically contributed virtually all of the greenhouse gases already in the atmosphere. The resolution opposed the draft treaty 'because of the disparity of treatment' between so-called Annex I Parties (primarily the United States, Canada, Japan, Australia and Europe) and developing nations and asserted that that the 'disparity of treatment' could 'result in serious harm to the United States economy, including significant job loss, trade disadvantages, increased energy and consumer costs, or any combination thereof'.[4]

Given this reluctance, a reluctance that is to some extent understandable particularly given the rate of growth in Chinese fossil energy use, it is not surprising that Kyoto was never ratified by the United States or that President Obama felt that he must proceed with great caution in Copenhagen. The question this chapter addresses is: can this extreme caution regarding climate change on the part of largest emitter of GHGs be reversed? More than that, is there any prospect that larger numbers of Americans might come to adopt a more cosmopolitan perspective and take up the obligations of wealthy nations to act first and decisively on climate change?

AMERICAN EXCEPTIONALISM AND COSMOPOLITAN ATTITUDES

The resistance to effective climate change action within the United States has many dimensions. Most obvious are the entrenched economic interests, particularly the oil, coal, utility and automobile industries (and many ancillary and related industries including fast foods, motels, rubber and steel). Some oil companies (most notably Exxon Mobil) have funded climate denial campaigns while others are somewhat more amenable to some forms of policy change. These industries as a group successfully resisted, until recently, enhanced government requirements for fuel efficiency in vehicles. Particularly successful was the lobbying effort that kept pick-up trucks and SUVs defined as outside the 'automobile' category and thereby exempt from fleet average fuel efficiency standards. Automobile companies then disproportionately promoted and sold those vehicles (on which both profits per vehicle and fuel use were highest).

Ironically, this effort was undermined by the war in Iraq, a military effort not unrelated to securing global oil supplies. The war in Iraq, as well as political instability elsewhere in the Middle East and in Nigeria, drove oil prices sharply higher beginning in 2003 and severely hurt vehicle sales, especially the sales of the larger vehicles favored by North American firms. Ironically again, the North American automobile industry had a near-death experience caused by, of all things, decades of effective lobbying. Had the industry invested money in new vehicle designs and technologies, and retooling their operations, rather than on lobbying to avoid the need for doing these things, they might well have lessened the financial difficulties they experienced. At the same time, some large corporations, including for example General Electric and FedEx, have taken a pro-climate change action position. Also, the automobile industry is now no longer financially autonomous and in no position to block government action on the issue.

Nonetheless, the American public is still exposed to relentless media-based anti-climate action campaigns, quite unlike that experienced anywhere else in the world. So-called talk radio in the United States is dominated by extreme conservatives almost all of whom deny the existence of global warming or any human role in it. Sean Hannity, who appears daily on both ABC radio and FOX television, for example, has asserted that 2009 was one of the coldest (or even 'the coldest') on record. In fact 2009 globally was, according to the US National Oceanic and Atmospheric Administration, one of the warmest.[5] The many Americans who depend primarily on conservative media sources for information are uninformed about the basic scientific facts of the issue and are indeed often distrustful regarding science generally. These misinformed views are also accepted or utilized by many who seek public office, including almost all of the Republicans who won contested seats in the United States Senate in 2010.[6] How can such misinformation be so influential when virtually all material published in scientific journals supports a consensus view that temperatures are rising and human-sourced GHGs have contributed to that shift? The pattern of misinformation begins with well-funded plausible doubts raised by scientists generally outside of the consensus view based on findings reported in scientific journals.[7]

As Paul Krugman put it, this research

> works for its sponsors, partly because it gets picked up by right-wing pundits, but mainly because it plays perfectly into the he-said-she-said conventions of 'balanced' journalism. A 2003 study . . . of reporting of global warming in major newspapers found that the majority of reports gave the skeptics – a few dozen people, many if not most receiving direct or indirect financial support from Exxon Mobil – roughly the same amount of attention as the scientific consensus, supported by independent researchers.[8]

A 2006 study by the *National Journal* found that many Washington Republicans, including many US Senators and their staffs as well as some who worked in the Bush administration, were more skeptical that climate change was real than were typical American citizens, and sharply more doubtful than the scientific community as a whole.[9] Some Republicans, such as the former governor of California Arnold Schwarzenegger, have advanced climate change action, but more recently most elected Republicans reject or override the conclusions of the scientific community. President Obama cannot get an international climate change treaty ratified unless at least some Republican Senators vote to ratify it and that seems less likely than it was prior to the 2010 midterm elections. Even prior to those elections Democrats abandoned the attempt to pass comprehensive climate change legislation.

Moreover, opposition to comprehensive action on climate change is broader than those that doubt that climate change is taking place or who doubt that it is caused by human activities. Many might allow that climate change is real but stand firmly against governmental interference in the market for almost any reason, let alone in order to mitigate impacts that may take many years to become fully visible and may end up impacting other nations (such as low-lying nations like Bangladesh or already dry nations like Ethiopia) more than they affect the United States. They brush aside as 'alarmist' concerns of scientists and policy analysts that it will take decades to change energy use patterns and by the time impacts are fully upon us we will not have time to make the necessary adjustments without considerable economic and social disruption.

Thus climate change remains a 'controversial' idea with more Americans than it does within many other nations. This, however, is not the primary reason that some Americans are slow to adopt cosmopolitan views on this, or any other issue, issue. A more important reason is a broad-based belief in American exceptionalism, the view that America is somehow especially well-suited, even destined to lead the world. A second reason is a parallel distrust on the part of many Americans of any semblance of global governance. Global governance is viewed by many as a means of undermining American power and American autonomy. A third reason is an entrenched trust in the use of military might as a means of resolving problems. Not all Americans, by any means, believe that America is qualitatively unique, a nation unlike all others and superior to them, but many do and they are not without influence. This view dates at least to the 19th century idea of manifest destiny which held that the United States was destined, perhaps even divinely ordained, to expand across North America (and later beyond). This view eventually became a singularly American mission to promote and defend democracy the world over – a view espoused by presidents as diverse as Theodore Roosevelt, Woodrow Wilson, Ronald Reagan and George W. Bush.

The mission began with making a nation that stretched from coast to coast and had, from the days of the Monroe Doctrine, special obligations with regard to, and privileges within, the Western Hemisphere. By the latter 20th century, the scope of American ambitions was increasingly global. By the beginning of the 21st century the United States was spending more on arms than all of its possible rivals and all of its allies *combined.* At the end of the Cold War, some Americans saw a potential peace dividend whereby some military spending could be channeled into long-neglected domestic needs including infrastructure, health care, improved schools and affordable post-secondary education. This hope, however, did not fully allow for the strength of the political interests and lobbying apparatus that had grown up around military expenditures.

The power and the ambitions of those interests were highly visible in the Project for a New American Century (PNAC) document, published in 2000 and entitled *Rebuilding America's Defenses*.[10] Those who signed the document became central figures in the Bush administration and included Paul Wolfowitz, Robert Kagan, Lewis Libby, Stephen Cambone and John R. Bolton. The statement sought nothing less than American military dominance of the world and saw this outcome as a logical consequence of the end of the Cold War. The report dwelled on what must be done to 'preserve American military preeminence' and to meet the 'requirements of a strategy of American global leadership'. The report in effect presumed the necessity of an American military presence in over 100 nations around the world and indeed celebrates this reality.

The report also made the following observation regarding the changes (and additional military spending) necessary to preserve military preeminence: 'Further, the process of transformation, even if it brings revolutionary change, is likely to be a long one, absent some catastrophic and catalyzing event – like a new Pearl Harbor.'[11] It is little wonder then that the Bush administration saw the attacks of September 11 as an opportunity to remake America's use of military power and American foreign policy as a whole. The view expressed in the PNAC document became the defining view of the Bush–Cheney administration and remains a perspective that has not been discredited among the American public as a whole. Military spending has not been reduced under Obama's leadership even if that had been the preference of some in his administration. This presumption of exceptionalism and belief that the world needs American global military preeminence, in effect, explains in part the great American ambivalence regarding global governance. Those that prefer global dominance are almost certain to see global governance as messy, inconvenient and untrustworthy in comparison. To come to this view American conservatives have had to reject much from America's historic approach to international affairs. Clearly there is a struggle going on over the central and defining outlook of the nation – a struggle that also encompasses the issue of climate change.

Following World War I the United States was instrumental in establishing the League of Nations; following World War II it made a concerted effort to create the United Nations and to locate its headquarters on American soil. More recently the United States was actively involved in putting in place any number of cooperative efforts with regard to global scale environmental problems. But all along there has been an undercurrent of distrust regarding these efforts, an undercurrent that was in the 1950s confined to extreme right-wing fringe elements such as the John Birch Society, which at the time hallucinated that the United Nations was poised to take over the United States. With the rise of neo-conservatism,

America's past internationalist efforts have come to be seen as 'unrealistic' or even quaint in some circles. Politically broad-based doubts regarding many forms of multilateralism are advanced in conservative media and are very much a part of the belief structure of many Americans. These views were a common perspective within the leadership of the administration of George W. Bush. Between 2001 and 2008 international treaties were voided in several areas including nuclear test bans, anti-ballistic missile defense, and US opposition intensified regarding the International Criminal Court and many other forms of multilateral action.

The Kyoto Treaty was explicitly rejected, and all succeeding efforts at regulating GHG emissions were turned back by the Bush government. Few, if any, new multilateral initiatives on any issue, other than seeking support for America's war efforts on America's terms, were sought. International laws regarding torture, access of prisoners to the International Red Cross and the United Nations Charter's provisions regarding the unprovoked invasion of other nations were questioned and in some cases ignored. Some of the institutions or rules that were put aside or circumvented had originally been initiated by the United States. Opposition to global governance is now widespread within American conservative circles. It is presumed by most American conservatives that the joint administration of global problems as diverse as the prosecution of terrorists or climate change is in effect a way to set limits on American power and to undermine American well-being. Even something as benign as a shared bicycle program in the City of Denver was attacked by Dan Maes, the unsuccessful Republican candidate for Governor of Colorado, as a worrisome imposition on Coloradans 'by an organization affiliated with the United Nations'.[12]

What Hofstader called the paranoid style in American politics thrives in the 21st Century.[13] Nightly to an audience of millions on FOX television Glenn Beck and others spin lurid and highly negative visions of the Obama government. At his most extreme Beck imagines the government confiscating America's guns or setting up camps for its domestic political enemies (presumably including those in his audience).[14] Hofstadter was originally writing about America in the 1950s and 1960s, but today America is very different: 1) cable news and conservative talk radio do not filter out the most extreme ideas, as newspapers and magazines and network television once did, but actively promote them; 2) there are fewer effective arbiters of the limits of extremism among conservative intellectuals; and 3) there are fewer elected moderates from New England or New York or elsewhere within the Republican legislative delegation – many of today's elected Republicans accommodate or embrace, rather than shy away from, the extremist Tea Party movement.

This increasingly broad-based neo-conservative thinking is inherently anti-cosmopolitan and stands foursquare against collective global action on climate change. Indeed it opposes all collective global action and thereby pairs well with explicit climate change denial, seeing the very idea of climate change as a machination of international, predominantly European, forces which are seen to be inherently socialist in outlook. Bush administration spokesmen asserted that climate change action was a threat to the American way of life. Some elected conservatives today have toned that down and merely argue that action on climate change, such as cap-and-trade legislation, would hurt economic recovery or act as a brake on economic growth. However, a significant proportion of the supporters of the Republican Party, especially the right-wing media and their faithful listeners, simply reject the science on this and other subjects in favor of 'faith-based' perspectives.

A wide cross-section of Americans of all political views, more than citizens of most other developed nations, believes in the efficacy of military action. Interestingly, excessive military spending is almost never seen in the United States as 'a threat to economic recovery' or 'a brake on economic growth'. It is widely seen to be precisely what government should be doing, for some almost the only thing it should be doing. Robert Kagan, an American neo-conservative foreign policy analyst made this observation on the eve of the 2006 US elections:

> Americans have more belief in the utility and even the justice of military action than do most other peoples in the world. The German Marshall Fund commissions an annual poll that asks Europeans and Americans, among other things, whether they agree with the following statement: 'Under some conditions, war is necessary to obtain justice.' Europeans disagree, and by a 2 to 1 margin. But Americans overwhelmingly support the idea that war may be necessary to obtain justice. Even this year, with disapproval of the Iraq war high, 78 percent of Americans agreed with the statement.[15]

The mindset that sees war as an acceptable activity is not unrelated to the United States never having fought an extended war against another nation on American soil. Over time this focus on the military dimensions of international affairs has led to the view that America is special and to seeing other nations as either enemies or weak and unwilling to fight or to spend sufficiently on defense. The view also blends well with a distrust of collective global initiatives and an abhorrence of both global governance and efforts that seek to build universal equality. Kagan made his point arguing that even if the Democrats gained in Congress, to a considerable degree they too saw the United States as 'the indispensable nation' and that a Republican defeat in 2006 would not mean the end of the war in Iraq or other aspects of Bush administration foreign policy. Generally speaking Kagan was right in this assessment and thereafter US troop numbers in Iraq were *increased* and have only now declined, with many being relocated to Afghanistan. Seeing through a reversal of Bush policies on climate change will not be easy either – the egalitarian principles inherent in a cosmopolitan approach to climate change are likely to send most Republicans into a rage and the administration will likely be especially wary of provoking an intense response with Republicans now in control of the House of Representatives.

Overall US polling support for the Obama administration remained favorable after one year and, while it was weaker by 2010, many Americans continue to reject conservative attitudes in favor of a far more cosmopolitan outlook. In short, America is a complicated nation characterized by a diverse array of opinions, a nation very much at odds about its future course. A majority of Americans still accept that climate change is real and favor policies to reduce GHG emissions, but that majority declined between April 2008 and October 2009.[16] In 2008, the nation elected a federal government that was committed to decisive action on the issue, but since that time it is clear that resistance is strong and building. Only a minority of Americans has even heard of cap and trade legislation and only 35 percent of Republicans believe that there is solid evidence of global warming. In broader terms, a majority of Americans came to oppose the occupation of Iraq. More than that, an overwhelming majority of Americans viewed the former Vice President (Richard B. Cheney) in negative terms, with a popularity rating consistently below 20 percent of the population. Cheney is not inclined to cosmopolitan views, to say the least. He and most neo-conservatives unrelentingly favor maximizing US national power and reject all forms of non-dominant international cooperation; but a majority of Americans reject many of his most extreme views.

Following the re-election of Bush and Cheney in 2004 vast numbers of Americans signed up to a campaign later published as a book of photos that visually captured individual Americans apologizing to the world for the re-election of the Republicans.[17] In contrast, in 2008 Republicans attempted to turn the international popularity of then-candidate Obama (following the massive public turnout for him in Germany) against the Democrats by suggesting that Obama's international appeal implied somehow that Americans should distrust him. That effort failed to gain majority support, as did Republican efforts to intimate that Obama was somehow foreign, but deep seeds of distrust were sown with that part of the Republican base prone to distrust all things foreign.[18] What is apparent in all of this is that there is a profound ambivalence in American culture and opinion. A significant minority of Americans are even prepared to believe that President Obama was not born in the United States and therefore is ineligible to be President. Many believed Sarah Palin when she asserted that health care reform would establish 'death panels' to disallow health care for elderly Americans. Doubting that climate change is real, or that the United States has an obligation to do something about it, would seem by comparison to be a more modest deviation from reality.

COSMOPOLITANISM AND CLIMATE CHANGE

The core challenge, then, for any American leader wishing to act effectively on climate change, is the instinctive distrust many Americans have regarding the importance of international cooperation or any need for individual Americans to change their everyday behavior to benefit future generations in other nations. The mindset inclined to American exceptionalism and hegemonic power sees other nations as either hostile to the United States (Muslim nations, China, Russia), socialistic (much of Europe) or perpetually seeking handouts from America (poor nations generally). This latter belief is particularly egregious and off the mark given that the United States contributes less per dollar of GNP to non-military foreign aid than do most other wealthy nations. Again, this neo-conservative perspective is not a majority view, but it is a view held all or in part by millions, and that permeates America's conservative media where President Obama was castigated for, of all things, promptly sending aid to Haiti following the devastating earthquake of 2010. This viewpoint is America at its worst, but it is at the same time a viewpoint that is very quick to remember America at is best – as it was in the establishment of the Marshall Plan following World War II or providing relief in many other global disasters. Few who hold to this point of view will accept special obligations with regard to climate change based on historic GHG emissions, a key component of any cosmopolitan approach to the issue. Denial of the very existence of climate change is far more likely, but even if that view is overcome by scientific evidence some may still believe that America simply deserves everything it has and more. It is as if, in this view, Americans are, because of their 'hard work and unique political and economic system', exempt from the laws of elemental justice as well as the chemistry and physics of climate science.

This body of opinion is even more influential than it might appear to be in polling data because the Republican Party and conservative media continuously imply that those who do not hold this view are less vigilant defenders of American interests and the American way of life. In this context many Americans are unmoved by the possibility of flooding in Bangladesh, loss of glacier runoff as a water supply in central Asia or Latin America or increasing periods of drought in Africa. Many do not know such things are coming, but even if they do the possibilities are seen as just more troubles in inevitably troubled places. It is this anti-cosmopolitan mentality held by many influential Americans that has thus far kept the United States from assuming a position of global leadership on climate change even though a modest majority might well welcome such action. This is ironic given that the environmental movement itself first took hold as a mass movement within the United States and still speaks for many Americans. Large US environmental organizations have multi-million dollar budgets supported by donations from individual Americans. In recent years, however, these organizations have had greater influence at the state and local than at the national level since the federal government has been and remains stymied on strong climate change action.

America is torn between those that are comprehensively cosmopolitan and embrace fairness to all individuals on a global scale and those that see the world as a sea of troubles for and threats to America. Being interested in those troubles other than through the lens of American interests is taken within the latter perspective to be evidence of weakness, and the United Nations itself, a driving force on climate change action, is seen by some to be an institution inherently hostile to American interests. Nonetheless, there are cosmopolitan forces within the United States that could be expanded and mobilized more effectively. One US-based campaign to mobilize on a global scale regarding the issue of climate change is 350.org, a group named after the parts per million (ppm) of carbon dioxide in the atmosphere deemed to be a safe level (and a level below the present level of about 388 ppm). In late 2010 this group, headed by author Bill McKibben, sponsored and coordinated more than 7,000 events in 188 countries.[19]

Other US-based cosmopolitan initiatives include global-scale micro-lending groups such as Kiva and large citizen-mobilizing organizations like Avaaz, but perhaps the most explicitly cosmopolitan undertaking may be the Great Transition Initiative (GTI), a group with global links but based at the Tellus Institute in Massachusetts.[20] GTI argues that a 'planetary phase of civilization is ready to crystallize' and that a global transition to a sustainable and just society has become a real possibility. These organizations generate less media attention than much smaller groups that seek to block the construction of mosques in New York or Tennessee, but they have the potential to mobilize individual and institutional actions (including those by firms and civil society organizations). Even when the President is unable to gain sufficient Senate support for new treaties or to pass climate change legislation these and other groups, or individuals or firms or local governments acting on their own, can initiate significant and highly visible undertakings.

President Obama must seek change at the national level within a context where his approval ratings are at best only modestly positive and his critics can block any new legislative initiative. Even if he cannot move Congress to take significant action on climate change, he can act through administrative initiatives and he can recognize and encourage a cosmopolitan citizen-based mobilization acting directly to counter climate change. Some non-cosmopolitans may then be brought along through witnessing exemplary changes – citizens, churches and city halls installing solar panels, firms producing more energy efficient appliances and selected state and local governments working to improve public transit. Moving forward in these ways will hopefully remind Americans of their better natures. It is also important to make it clear that shifting the nation's energy economy is crucial to restoring prosperity and rebuilding America's industrial economy and will also help to make it clear to Americans that energy security is crucial to national security.

Americans have a dark side without a doubt, but many, including conservative Americans, have a better side. Many are generous with their money and skills. They volunteer to help the poor and are exceeding generous with personal donations to charities the world over. The Peace Corps captured this ethos of sharing as did AIDS funding in Africa under President George W. Bush. As noted, modern environmentalism itself had American origins as did the United Nations. Even America's recent history is far from a one-sided rejection of cosmopolitan values. It remains possible that a majority of Americans could come to support American leadership on global climate change.

Moreover, it is simply not the case that climate action is all costs and no gains. Energy supply options like wind, geothermal and solar expand the job market, and economic interests develop that will push for further expansion. Renewable energy, energy efficiency and public transit are all more labor intensive than the economic activities with which they compete (a finding that was clearly demonstrated in economic analyses dating back to the energy crisis of the late 1970s and early 1980s).[21] Environmental organizations that have made the case for the economic benefits of climate change action include the Apollo Alliance of environmentalists, community groups and labor unions as well as the noted environmental activist Van Jones.[22] At a time when much of the US midwest has been de-industrialized as manufacturing has shifted to China and elsewhere, many green energy jobs would be certain to stay in North America. Insulation can only be installed where buildings already exist, transit must be installed, operated and maintained at point of use and cities will be rebuilt more compactly when public transit is developed.[23] Even if some components are manufactured abroad, renewable energy's installation and maintenance jobs are inevitably at the point of operation.

A case can be made that resolving climate change is also in America's interest from a national security standpoint. Climate change is likely to be profoundly destabilizing. Also, excessive dependence on oil imports adds to the pressures for the US to be involved abroad in places where it otherwise might have little or no strategic interest. And the outflow of money for oil imports undermines US economic security and in some cases may indirectly provide funds to actual enemies including terrorist sympathizers. As noted, this perspective has not been lost on strategic thinkers within the US military. These aspects of climate change are important because they could help to offset claims from neo-conservatives that climate change action threatens the economy, a sign of weakness and cooperation with 'a global agenda' on the part of the administration. In the contemporary mindset of some Americans it may be easier to sell policies in terms of national security and prosperity than in terms of doing what is right.

Cosmopolitan assertions regarding fairness related to climate change will appeal to many Americans, who can be convinced to take individual and community actions, but they will still be resisted by others. It is important to argue that everyone on the planet has a right to be protected against the risk of climate change, but it may not be enough in the first instance to get support for dramatic American leadership on the issue from a large majority of Americans. The problem is not so much that a majority could not be convinced, but that the forces of conservatism in America again seem to have an edge and have a number of institutional advantages on their side.

INSTITUTIONAL BARRIERS TO CLIMATE ACTION IN THE UNITED STATES

The challenge of climate change for the United States is formidable not just because of the habit of hegemony and the resulting perspective of Americans. There are also structural and institutional barriers, some rooted in the US Constitution, some in governmental procedures and habits and some rooted in the nation's political economy. Increasingly, the greatest barrier to effective action on any policy issue is the United States Senate, both constitutionally and procedurally. Constitutionally, each state chooses two senators regardless of the population of the state. Large urban, high population states like New York, Illinois and California, states that in recent years have leaned toward progressive politics, each have two. So do low population, rural states that lean towards the conservatives like Wyoming, Idaho and Alaska. Conservatism is also entrenched in most of the South and many of those states have relatively small populations.

The result is that a minority of Americans, a mostly rural and Southern minority, can elect a substantial proportion of the Senate. There is also a curious Senate procedural rule that has evolved to the point where a 60-vote majority in the 100-seat Senate is needed to bring legislation to a vote, to ratify a President's administrative appointments or to ratify a treaty. Until recently this requirement was only rarely utilized; now it would appear that it is brought to bear in virtually every instance. As was made plain in the year-long debate over health care, this procedural rule placed enormous power in the hands of the Senators with the deciding votes during the first two year of the Obama administration. From 2008 to 2010, these were the most conservative one or two Democratic senators and the most liberal one or two Republican senators. This procedure and practice, combined with the disproportionate representation of conservative states and heavy corporate funding of election campaigns, results in a strong advantage to entrenched interests and the status quo. In addition, the limits on corporate financing of electoral politics were effectively removed by a Supreme Court decision in January 2010 (in *Citizens United v. Federal Election Commission*). This decision may contribute to skewing climate change policy towards the interests of energy and related corporations, although another Court decision (*Massachusetts v. Environmental Protection Agency*) in 2007 empowers the EPA to regulate carbon dioxide emissions without additional Congressional action.

Control of the Supreme Court currently rests with key appointees of George W. Bush and the next retirements to come are likely to be from the liberal side of the Court. Thus the court's composition will remain static for years to come. The Supreme Court is another constitutional structure designed to prevent democratic excesses, a task in which the document has succeeded brilliantly. Curiously, the Citizens United ruling gives greater power to corporate campaign donations in the name of *the free speech provisions of the Constitution*, as if corporations were citizens like any individual and as if concentrated wealth could pose no threat to democratic functioning or the opportunity for individual citizens to be heard. As Mr. Justice Stevens incisively asserted in his dissenting opinion: 'While American democracy is imperfect, few outside the majority of this Court would have thought its flaws included a dearth of corporate money in politics.'[24] One reason that corporate money is so decisive in American politics is the combination of separation of powers, the aforementioned equal allocation of Senate seats by states and weak political party discipline. National-scale corporate money goes a very long way in sparsely populated states like Montana and Arkansas and senators from those states are, regardless of party, very wary of challenging important economic interests.

Thus the US political system renders decisive steps extremely difficult and transformational politics almost impossible (except in extreme circumstances such as war, depression or following an event like September 11). Climate change, however, has not risen to that level of concern within the United States – its threat is inherently chronologically down the road and its impacts are uncertain as to their timing and detail.

Also difficult politically is that the energy transformation necessary to avoid climate change is expensive and must precede the worst impacts by several decades. It is not easy to bring people to understand these realities when the media prefer to cover policy issues that have greater immediacy and stories which have a sensational dimension like the saga of Tiger Woods. A *potential* climate crisis, especially one that is likely to affect other countries first, is not the sort of threat that easily arouses the media. Without that attention it is difficult indeed to rally all the separate elements of American government sufficiently to accomplish decisive action. The US governing system is arguably better suited to the slower-moving 18th century when it was created than to the globally integrated, rapidly paced 21st wherein the United States is the dominant military and economic power.

The system has evolved toward greater efficiency over the centuries, but in recent decades it seems to have lapsed back into a debilitating incapability for taking decisive non-military action of any kind. There is more power in the hands of the President than in the original Constitutional design, but that power is primarily in the realm of strategic and military policy. Climate change has an unavoidable domestic policy component and there the resistance to change is deeply entrenched. The President could not see through decisive legislative changes regarding climate change. Indeed that was why Kyoto ratification was impossible in the waning years of the Clinton administration and why Obama, in his 2010 State of the Union Address, had to preemptively offer concessions regarding offshore oil drilling and 'clean coal' as a part of an 'energy and climate package'.

At the same time that the American political system has serious problems regarding the climate issue, there are some structural elements within the system that could help. First and foremost among these is federalism. Historically, when the White House and the federal government have been hostile to environmental policy action, selective state and local governments have stepped up to take initiatives on their own. This was very much the case when Ronald Reagan was President and sought to undermine the enforcement of environmental regulations. When he did that, progressive states, especially California, established effective environmental rules within their administrative space. Cosmopolitan Americans can again act effectively at this level and as individuals and as investors even if the national government is legislatively stymied.

During the Reagan years US environmental organizations grew rapidly in both membership and financial contributions.[25] The Sierra Club and numerous other organizations expanded by mobilizing concern that existing environmental protections, which at the time were as good as any in the world, would be undermined. Reagan opted not to repeal environmental legislation; he chose rather to prevent the enforcement of existing laws through administrative means (by, for example, appointing noted opponents of environmental protection and cutting budgets for inspections of industries by government officials). Obama could in effect do the opposite by encouraging cosmopolitan initiatives on climate change by civil society as well as industry and local governments.

State and local climate change environmental initiatives also actually increased during the George W. Bush administration. At the municipal level many undertakings started when the Kyoto Protocol came into force outside the United States. Mayor Greg Nickels of Seattle challenged mayors elsewhere to join the effort to reduce global warming gases. Mayors from San Francisco, Portland, Minneapolis, Burlington, Vermont and Boulder, Colorado joined him and they collectively wrote to 400 other mayors. The US Conference of Mayors backed the effort unanimously. By 2006 it had the support of 250 American city governments, by 2008, more than 600 with a total population of 67,000,000 and in 2010 more than 1,000 American cities had signed up.[26] Seattle's climate change efforts are handled by an Office of Sustainability and the Environment and include urban reforestation, enhanced bicycling opportunities, green roofs on city buildings, improved walkability initiatives, technical assistance to builders regarding energy efficiency, zoning changes downtown that discourage sprawl, reduced use of fossil fuels in city-owned vehicles and increases in recycling and composting. In short the full array of policies over which cities have a major voice. More recently, Mayor Bloomberg of New York City has become an innovator on climate change as well by, for example, working to transform the city's massive taxi fleet to hybrid vehicles. Some states also acted particularly effectively with California, New York and New England leading the way.

All of these efforts have been pushed by cosmopolitan-oriented citizens acting on the local level. The results were significant – despite eight years of federal resistance to climate change commitments GHG outputs in the United States grew more slowly than they did in, for example, Canada which signed and ratified Kyoto. Even a conservative state like Texas established significant tax incentives for the development of wind energy. In short, if the US government cannot act on environmental matters, governments at the sub-national level have more than once acted more forcefully than they otherwise might have. This has great potential as part of any US initiative on climate change. Should the Obama administration be unable to act decisively on this issue, a more decentralized action plan could evolve.

Decentralized climate change action could gain ground over time because large corporations produce goods for national and global markets and therefore prefer to deal with a unified set of regulations rather than rules that vary from state to state. Once several large US states have established rules that cannot be reversed, those corporations are just as happy to see uniform national requirements. Their political influence could tip the balance if Washington is mired in political gridlock. Indeed on the issue of climate change while some leading oil companies promote climate change denial, other corporations feel they can do better in a policy climate that promotes energy efficiency and renewable energy through government incentives and disincentives. Also in support of government action are firms such as the manufacturers of public transit vehicles, insulation and solar panels that benefit directly from such policies.

Thus within the United States political support for climate change action might come less from collective global arrangements than from the political and economic appeal of incentives that empower firms, individuals and states to act. The Texas tax incentives for wind energy are one example of this potential. In Canada, as well, the government of Ontario has an incentive system (patterned after German arrangements) that rewards homeowners by paying a premium for selling clean power into the provincial grid. This approach might appeal to many Americans. Decentralized policies work not only to reduce GHGs, but to mobilize new firms and employees to have a vested interest in climate change prevention. This grassroots dynamic would be politically important within the United States. Many conservative Americans are aggrieved that the nation's manufacturing economy is slipping away. There is massive underemployment of industrial and construction skills and an intense desire for industrial jobs. More than that, this approach undercuts the forces of resistance to climate change action. They cannot believably argue that the federal government is 'forcing people to do something' and/or 'acting at the behest of foreign interests'. The effort is rather a matter of 'putting Americans back to work' and 'empowering local, individual and corporate initiatives'.

When much of a nation's media and one of only two viable national parties resist any and all change, framing policies is critical to achieving new initiatives of any kind. President Obama sought to counter climate change skepticism when he framed his call for climate change action in his 2010 State of the Union address in terms of economic growth and even nationalism. As he put it:

> I know that there are those who disagree with the overwhelming scientific evidence on climate change. But here's the thing – even if you doubt the evidence, providing incentives for energy-efficiency and clean energy is the right thing to do for our own future – because the nation that leads the clean energy economy will be the nation that leads the global economy. And America must be that nation.[27]

This is decidedly not a cosmopolitan appeal, but it could complement such appeals.

THE CENTRAL CHALLENGE TO COSMOPOLITANISM

A desire for effective climate change action begins for many in a cosmopolitan perspective, but initiatives grounded in cosmopolitan thinking can be linked to local needs in ways that encourage some with less than fully formed cosmopolitan inclinations and even some climate skeptics. In the United States convincing those that are not fully cosmopolitan in their views may be especially important because within the American political system substantial minorities can readily undermine policy initiatives. Successful climate change action requires more than legislation – it needs a broad-based buy in and the active participation of home owners, consumers and business people, indeed anyone who actively participates in the economy. Widespread participation in climate change solutions is essential; governments cannot simply sign a treaty and be done with it. Especially in nations where per capita energy use is very high, large majorities need to adopt new transportation habits, insulate their homes, use energy efficient appliances, encourage their utilities to supply electricity from renewable sources and reduce their consumption of energy intensive products and services like air travel. Governments can encourage changed behavior through price incentives, but ultimately the changes will come from the changed behavior of firms, municipalities and individuals. Crucially those with cosmopolitan outlooks can be early adopters prior to policy action or in the face of policy gridlock.

A high proportion of citizens must understand both the scale of the task and the reasons why that task must fall disproportionately on the wealthiest nations, but other motivations may be necessary to reduce resistance and to broaden participation. That cosmopolitan understanding is out of keeping with a central piece of conventional wisdom that is central to the 'American way of life'. That bit of conventional wisdom is this: individuals can acquire whatever goods and services they both desire and can afford to pay for. If that rule remains in place without limits the problem of climate change probably cannot be resolved short of radical increases in the price of fossil fuels – unless Americans can be brought to make different choices than they have made in the past regarding comfort, convenience and status. Such things will not be easily sold because the GHG emissions of each human on the earth must (on average) decline even while some people's emissions actually rise. Millions in China or India are going to move into larger living spaces or buy refrigerators or cars or cell phones that need recharging. Most of the previous generation never had such luxuries. Historically, the average per capita emissions of GHGs in these countries are almost negligible compared to the average in North America or Europe. There is no rational argument within any semblance of a cosmopolitan framework that would justify locking people in these nations into the low level of emissions they have had in the past when they are now able to attain some of the goods and services we in the rich nations have had for a century.

The central challenge of climate change policy is inherent in the degree to which historic and current emissions are disproportional between rich nations and poor. Total global emissions must be reduced and if current emissions are distributed equitably, and historic emissions are taken into account at all, the emissions of heavy emitters must be *greatly* reduced while the per capita emissions of some nations rise. This conclusion challenges the American ethos in two ways: it undermines the presumption that individuals with money should have what they want so long as it is legal and the presumption that the United States as a nation deserves every advantage it has. A cosmopolitan perspective sees the matter differently and only some Americans will accept that readily.

Vanderheiden (see Chapter 2), in his discussion of the ethical dimensions of establishing global climate change obligations, identifies two methods for determining the obligations of historic low per capita emitters.[28] He distinguishes between levels of per capita GHG emissions – with the line between the two determined in one of two ways (and emissions reduction obligations falling only on those emissions over the line). One way is to determine a safe (absorbable) level of total historic emissions per capita and allocate that amount globally to nations based on population. A second method calculates the basic minimum a person would be:

> allowed to emit in order to meet their basic human needs and to which all persons would therefore be entitled as a matter of basic rights, even if the

world's population producing GHGs at this rate would still contribute to the increasing atmospheric concentrations that cause climate change.[29]

There are two core problems in establishing these options. Some Americans will balk at having to reduce emissions substantially and will need others to set visible personal examples. It will also be very difficult for high emitter nations, especially the United States, to reduce their emissions sufficiently if emissions from countries like China actually continue to rise as justice, and Vanderheiden's second method, suggest they should be permitted to do. Clearly, however, historic emissions are not sufficient as a basis for determining allowable emissions if they result in a formula which dooms the planet, because this may well happen given the possible warranted rate of increase of emissions in fast-growing economies with historically low emissions. In the end compromise and learning are unavoidable and must take place at a rate more rapid than might be expected for such a fundamental transformation. Perhaps the best prospect for genuine compromise lies in a genuine USA–China dialogue – a meeting of minds from the reluctant giants on the issue. Ultimately though that coming together will only succeed if the United States can find a way to take the first steps.

CONCLUSION: LEADING AMERICA TOWARD COSMOPOLITANISM

While there is significant resistance to cosmopolitanism within the United States, this can be turned around. The resistance is rooted in the long-entrenched habits and attitudes of hegemony. Some Americans feel superior to people in other nations, some smugly so; some simply feel blessed by good fortune. Many also believe that their nation has a proverbial, and real, target on its back. The paranoid style in American politics has long been there and terrorist attacks have added to it. Yet many Americans now realize that neo-conservatism mired their nation in low global esteem and undermined economic well-being to the point where American leadership is on the wane and militarily-based hegemonic power may no longer be affordable. America is clearly at a political crossroads.

The issue of climate change can be part of a transition to a post-hegemonic era in which the United States restores its leadership within a multilateral global governance system. Obviously, the world and the United States are presently a long way from that possible future and when put in those terms the idea might well be rejected outright by a majority of today's Americans. It is possible however that other forces and needs will push America in that direction despite the political predilections of conservative Americans. President Obama and many around him, as well as those in the corporate world, realize that the economic future, with or without climate change, lies with alternatives to fossil fuel dependence. In a sense, that realization is part and parcel of an outlook that is both cosmopolitan and realistic. The United States can no longer afford to simultaneously buy foreign oil in unlimited quantities and export industrial jobs at today's pace.

The energy and infrastructure alternatives necessary to resolve climate change can help to overcome Americans' resistance to economic and political adaptation. The necessary changes can be advanced in decentralized ways using predominantly market-based tools with cosmopolitan early adopters leading the way as the economic and political systems slowly shift gears. The changes can truthfully be sold politically as essential to economic recovery, to correcting imbalances in the American economy and to making it less urgent for the United States to be militarily active in so many unstable corners of the world. Most Americans would celebrate those changes. Cosmopolitan Americans will see that much more will be achieved in the process.

NOTES

1. John M. Broder, 'Climate Change Seen as Threat to U.S. Security' (9 August 2009), available at www.nytimes.com (accessed 15 January 2010).
2. Gwynne Dyer, *Climate Wars* (Toronto: Vintage Canada, 2009).
3. William J. Broad, 'C.I.A. Is Sharing Data with Climate Scientists' (4 January 2010), available at www.nytimes.com (accessed 5 January 2010).
4. Senate Resolution as quoted in Henrik Selin and Stacy D. VanDeveer, 'Global Climate Change: Kyoto and Beyond', in Norman J. Vig and Michael E. Kraft, *Environmental Policy: New Directions in the Twenty-First Century* (Washington, DC: Congressional Quarterly Press, 2010), p. 275.
5. See 'Quick Fact: Hannity Again Wrong on 2009 Temps in Attack on Global Warming' (1 December 2009), available at www.mediamatters.org (accessed 2 December 2009).
6. See Editorial, 'In Climate Denial, Again' (17 October 2010), available at www.nytimes.com (accessed 18 October 2010).
7. For an analysis of that consensus see Naomi Oreskes, 'The Scientific Consensus on Climate Change', *Science* (1 April 2006): 1686.
8. Paul Krugman, 'Enemy of the Planet', available at www.nytimes.com (accessed 18 April 2006).
9. See 'Inside Washington: Congressional Insiders' Poll', *National Journal* (1 April 2006): 5–6.
10. Project for a New American Century, *Rebuilding America's Defenses: Strategy, Forces and Resources for a New Century* (Washington, DC: Project for a New American Century, 2000).
11. *Ibid.*, p. 51.
12. See Christopher N. Osher, 'Bike Agenda Spins Cities Toward U.N. Control, Maes Warns' (4 August 2010), available at www.denverpost.com (accessed 6 August 2010).

13. See Richard Hofstadter and Sean Wilentz, *The Paranoid Style in American Politics* (New York: Vintage, 2008).
14. The latter claim was later withdrawn by Beck, but the seed is still planted. Beck has also asserted that the administration is riddled with socialists and, if one can judge from the signs at rallies that Beck has promoted on the air, Beck's listeners seem to have concluded that Obama's health care plans are similar to those of Nazi Germany.
15. Robert Kagan, 'Staying the Course, Win or Lose' (2 November 2006), available at www.washingtonpost.com (accessed 30 January 2011).
16. For opinion data on this issue see Pew Research Center, *Fewer Americans See Solid Evidence of Global Warming*, available at www.people-press.org/report/556/global-warming (22 October 2009) (accessed 30 January 2011).
17. James Zetlen and Ted Rall (eds) *Sorry Everybody: An Apology to the People of the World for the Re-Election of George W. Bush* (Irvington, New York: Hylas Publishing, 2005).
18. A recent poll of Republican opinion found, for example, that 63 percent of Republican voters believed that President Obama was a socialist and an additional 16 percent were not sure if he was. See www.dailykos.com/statepoll/2010/1/31/US/437 (accessed 30 January 2011).
19. See www.350.org (accessed 30 January 2011).
20. See www.kiva.org, www.avaaz.org and www.gtinitiative.org (accessed 30 January 2011).
21. See Robert Paehlke, 'Work in a Sustainable Society', in Roger Keil, David V.J. Bell, Peter Penz and Leesa Fawcett (eds) *Political Ecology: Global and Local* (London: Routledge, 1998), pp. 272–291.
22. Van Jones, *The Green Collar Economy* (New York: HarperOne, 2008).
23. See Peter Newman and Jeffrey Kenworthy, *Sustainability and Cities* (Washington: Island Press, 1999).
24. The court's decision in this case, including the dissents, is available at www.supremecourtus.gov/opinions/09pdf/08-205.pdf (accessed 30 January 2011).
25. See, for example, James A. Tober, *Wildlife and the Public Interest* (New York: Praeger, 1989), p. 38.
26. See www.usmayors.org/climateprotection (accessed 30 January 2011).
27. The text of the speech is available from www.whitehouse.gov (accessed 30 January 2011).
28. Steve Vanderheiden, *Atmospheric Justice: A Political Theory of Climate Change* (New York: Oxford University Press, 2008).
29. *Ibid.*, p. 72.

8. Overcoming the planetary prisoners' dilemma: cosmopolitan ethos and pluralist cooperation

Philip S. Golub and Jean-Paul Maréchal

INTRODUCTION

In a November 2008 article entitled 'Science: The Coming Century', Martin Rees writes that if the science of climate change is intricate, it is 'straightforward compared to the economics and politics'.[1] Indeed, global warming poses a unique nexus of economic, political and philosophical challenges – a 'perfect moral storm'[2] – for at least three reasons. First, its causes are globally dispersed and its effects are non-localized. Simply put, driving a car in Paris has no more effect in France than in Hong Kong, and vice versa. Second, the mean lifetime of fossil fuel CO_2 ranging between 30,000–35,000 years, there is a very long time lag before the natural carbon cycle neutralizes the anthropogenic emission of greenhouse gases (GHG).[3] The effects on sea levels of driving a car today will become manifest only in a couple of decades. Consequently, we have to consider the problem of justice both on the intragenerational level (between individuals, nations and social groups today) and on the intergenerational one (between people and societies living in different periods of time). Last but not least, our 'theoretical ineptitude' makes it difficult to solve the problems at hand. As Stephen Gardiner aptly points out, 'we are extremely ill-equipped to deal with many problems characteristic of the long-term future. Even our best theories face basic and often severe difficulties addressing basic issues such as scientific uncertainty, intergenerational equity, contingent persons, nonhuman animals and nature.'[4]

The facts on which our judgments should be based are well known and well established. Expressed as a global average, surface temperatures have increased by more than 0.70°C over the past century. The rate of temperature change has also accelerated. During the last 50 years the rate of increase was 0.13°C per decade, or twice what had been observed during the previous century, and has accelerated over the past two decades.[5] The 'linear' consequences of these temperature changes include the decline in snow pack and ice cap coverage, the retreat of glaciers, increasingly frequent extreme wet and dry weather events, the proliferation of pathogenic agents and so on. 'Non-linear' consequences include, for instance, the disturbance of deep-ocean circulation or abrupt collapses of ecosystems. The third IPCC report (2001) leaves no doubt about the causes of these modifications: 'There is new and stronger evidence that most of the warming observed over the last 50 years is attributable to human activities.' In fact, it is caused by the emissions of gases such as carbon dioxide and methane gases, to cite the most important ones, which are included in the 'GHG' category. Carbon dioxide concentration, which has risen 20 parts per million (ppm) during the last 8,000 years, rose from 280 to 379 ppm during the twentieth century. Atmospheric carbon dioxide and methane concentrations are now higher than at any time during the last 650,000 years.

These transformations threaten the human prospect. In the absence of decisions today, the effects sketched above will inhibit human development and possibly generate catastrophic systemic outcomes in future. By upsetting the balance of the ecosystem, anthropogenic interference with the climate[6] threatens the 'mother of all public goods'.[7] Yet, since there is no simple coincidence between the descriptive and the normative, we face major difficulties in translating knowledge into an effective global agenda. Having a grip on the facts does not tell us what should be done. James Garvey summarizes the complexity of the situation when he writes:

> The spatial and temporal smearing of actions and agency can be deeply confusing, because sometimes moral responsibility depends conceptually on another sort of responsibility: causal responsibility . . . We lack both the wisdom and the theory to cope with [global climate change]. It's possible . . . that our theoretical failure can lead to a moral failure, a kind of deception in which we focus on one part of the problem and not others. The complexity can be an excuse, a problematic excuse, for doing nothing at all.[8]

Indeed, the broad scientific consensus over global warming has not been matched by a shared commitment at the political level to find solutions founded on the general human interest.

The universal dimension of the crisis, a word whose etymology implies that we are facing a historical moment of *decision*, requires that we pay careful attention to the different issues raised by climate change. The aim of this chapter is to contribute to the theoretical debate by focusing on the issue of distributional justice among states and social classes. Nation states have varying capabilities and different historical responsibilities in the process of global warming, a reality that derives from their historical pattern of insertion at the centre or periphery of the global political economy, with implications regarding burden sharing. This issue featured prominently in the failure of the Copenhagen Summit of December 2009 to establish the foundations of a new binding global climate regime. At the same time, given the rise of a large class of consumers in 'emerging countries', a serious appraisal must imperatively distinguish between social classes. The purpose of new theorizing is to find ways to overcome the contradiction between global human needs and national rivalries due to the segmentation of the international system into discreet national units. To discuss these issues we have chosen to take the cases of the United States and China, the two largest emitters, the actions or non-actions of which will largely determine global outcomes. The question is how trust-building mechanisms could be developed that would lead to effective cooperation allowing states to break out of the prisoners' dilemma they otherwise face, and move forward towards a global project of protection of the atmospheric commons.

The first part of this chapter is devoted to structural aspects of the decision-making framework generated by American and Chinese GHG emissions. The second part examines fundamental debates concerning burden sharing between the US and China in global warming mitigation. The third deals with the intellectual and normative challenge to imagine new forms of ordered pluralist cooperation leading to convergence around common agendas that are in the overall human interest.

THE TRAP

There is a significant difference between knowing that something should be done and knowing what must be done. In the case of global warming, this difficulty is not only linked to technological challenges that are real and perhaps not solvable, but also to the problems raised by existing decision-making frameworks. There is a manifest contradiction between the understanding that survival in the long run depends on stopping the present course of climate change and the utility maximizing logic of *homo economicus*. The available scientific data leads to the inescapable conclusion that there is an overwhelming common interest to act today. Yet, at the same time, aiming to maximize short-term benefits, individual agents have an interest in continuing voracious consumption of fossil fuels and other natural resources. The logic of *homo economicus* leads all actors – states, firms or individual citizens, rich and poor people alike – to being unreasonable at the deepest level of reason. Economists call this a 'prisoners' dilemma' (PD). In the case of global warming, this dilemma is made particularly acute by three specificities: first, it concerns a 'common resource', second it is both intra and intergenerational and, third, it involves different categories of nation states with different historical responsibilities in global warming, including newly industrialized countries such as China.

Public Goods

The common resource endangered by global warming is the carbon-absorbing properties of the planet, properties that are sometimes called 'carbon sinks'. In what way are these carbon sinks a common resource? Goods can be classified according two attributes: whether they are excludable (people can be excluded from consuming them) and whether they are rival (one person's consumption of a considered good reduces the amount available to other consumers). Private goods (cars, oranges, and so on) are both excludable and rival. Public goods, on the other hand, are goods that are non-excludable (nobody can be excluded from consuming them) and non-rival (the consumption of one unit of such a good does not diminish another agent's possibility to consume it). Some public goods such as sunshine are produced by nature and virtually imperishable (at least on a human time scale). Others are man made, for instance national defense or lighthouses. If private goods can be efficiently produced and consumed on a competitive market, this is not the case for public goods. Because of the lack of financial incentives to produce them or to pay for their consumption (a free-rider problem) they must be provided by government and paid for with taxes.

There is a large spectrum of goods between purely private and purely public goods. These in-between cases – sometimes called 'impure' public goods[9] – belong either to the category of 'club goods' if they are exclusive but non-rival (for example, coded TV broadcast) or to the category of 'common resources' (or simply 'commons') if they are non-exclusive but rival (such as fish stocks). The climate can be considered as a pure natural public good or, to be more precise, it was a pure natural global public good prior to the ability of human beings to modify it. Since the end of the eighteenth century, when humanity became a kind of geophysical force (Vernadsky) and the planet entered in a new age that Will Steffen, Paul Crutzen and John McNeill call the 'Anthropocene',[10] the earth's climate has gradually became a global public good that has, in part, to be created by human beings. Of course, what has to be created is not the earth climate *per se* but earth climate stability.[11] Put differently, climate stability is no longer a 'pure natural public good', nor a 'pure artificial public good'. It is partially both. It is a 'global common' or, to be more precise, 'carbon sinks' are global common resources. The rise of earth temperatures – that can be defined as a 'public bad'[12] – is the result of an overuse by some economic agents (states, producers, consumers . . .) of the carbon absorbing capacities of the planet.

Left to market mechanisms alone, common resources suffer from overuse. A consumer can freely deplete the available amount of commons. This is what Garrett Hardin in a famous article called the 'Tragedy of the Commons'.[13] Thus, it can be said that the carbon-absorbing properties of the planet is a common resource and that GHG emissions are a form of appropriation of this common resource. The overuse of carbon sinks can be analyzed as a possible payoff of a prisoners' dilemma. The prisoners' dilemma generated by global warming is both intra and intergenerational.

An Intra and Intergenerational Prisoners' Dilemma

The intragenerational aspect of the dilemma can be expressed the following way:

(PD1) It is *collectively rational* to cooperate and restrict overall pollution: each agent prefers the outcome produced by everyone restricting their individual pollution over the outcome produced by no one doing so.

(PD2) It is *individually rational* not to restrict one's own pollution: when each agent has the power to decide whether or not she will restrict her pollution, each (rationally) prefers not to do so, whatever the others do.[14]

This is a good example of the traditional prisoners' dilemma. In the real world it could be resolved when the parties involved benefit from a wider context of interaction, that is to say when reciprocity or mutually beneficial decisions play an important role. Trade and security are good examples of such a situation. That is why Joseph Stiglitz has proposed to use trade sanctions as a mean of enforcing US participation in the Kyoto Protocol. In 2006, Stiglitz wrote:

> Fortunately we have an international trade framework that can be used to force states that inflict harm to others to behave in a better fashion. Except in certain limited situations (like agriculture) the WTO [World Trade Organization] does not allow subsidies – obviously, if some country subsidizes its firms, the playing field is not level . . . Not paying the cost of damage to the environment is a subsidy, just as not paying the full costs of worker would be. In most of the developed countries of the world today, firms are paying the cost of pollution

> to the global environment, in the form of taxes imposed on coal, oil, and gas. But American firms are being subsidized – and massively so. There is a simple remedy: others should prohibit the importation of American goods produced using energy intensive technologies, or, at least impose a high tax on them, to offset the subsidy that those goods currently are receiving.[15]

A slightly different line of reasoning leads to very similar conclusions. Game theory teaches us that to progress from rivalry to cooperation there is no need for friendship (fortunately: if there were, we should consider moving at once to another planet!) but only interaction taking place over time. To put it more precisely, sustainable relations between players may generate cooperation.[16] Who could deny that negotiations on climate control are an example of this kind of situation? But there is always the possibility of tit for tat: one player failing to attend might push others into retaliating by staying away. This outcome would, in the case of climate negotiations, be a recipe for disaster.

A disaster could also result from the intergenerational dimension of the prisoners' dilemma. The risk linked to this temporal characteristic appears clearly when we consider a 'pure' version of the intergenerational problem, that is to say a case where different generations do not overlap. This 'Pure Intergenerational Problem' (PIP) can be expressed as follows:

(PIP1) It is *collectively rational* for most generations to cooperate, (almost) every generation prefers the outcome produced by everyone restricting pollution over the outcome produced by everyone overpolluting.

(PIP2) It is *individually rational* for all generations not to cooperate: when each generation has the power to decide whether or not it will overpollute, each generation (rationally) prefers to overpollute, whatever the others do.[17]

This second trap (PIP) is worse than the first (PD). PIP1 is worse than PD1 because the first generation is not taken into account in the logical process. Thus, the first generation has no incentive to act and this inaction has a domino effect on subsequent generations, an effect that undermines the possibility of a collective project of compliance. PIP2 is worse than PD2 because, since the different generations do not coexist, they are unable to influence each other's behavior through the building of institutions setting binding norms. PIP is thus more difficult to resolve than the PD 'because the standard solutions to the Prisoners' Dilemma are unavailable: one cannot appeal to a wider context of mutually beneficial interaction, nor to the usual notion of reciprocity'.[18]

The Challenge of Emerging Countries

The dilemma has been sharpened during the last few decades by the impressive economic growth of some newly industrialized or re-industrializing countries. China is the perfect example of this new situation. Some figures give a very precise idea of the problems that are before us. In 1970, total greenhouse gas emissions were 23.9 $GtCO_2eq$. The share of the 'BRIC countries' (Brazil, Russia, India, China) was 5.9 (i.e. 24.6 percent) and the share of the OECD counties was 13.7 (i.e. 57.3 percent). In 2005, total greenhouse gas emissions were 46.9 $GtCO_2eq$. The share of the BRIC countries was 16.1 (i.e. 34.3 percent) and the share of the OECD countries was 18.7 (i.e. 39.8 percent). If no new policy actions are taken, in other words if we follow a business as usual scenario, global greenhouse gas emissions are projected to reach 71.4 $GtCO_2eq$ in 2050. The share of the BRIC countries will reach 26.2 (i.e. 36.6 percent), and 23.5 (i.e. 32.9 percent) for the OECD countries (see Table 8.1).

In 1980, Chinese emissions of CO_2 were about 1.4 billion tons, the emissions per capita being 1.4 tons. The American figures were respectively 4.6 billion and 20.5. In 1990 these figures were 2.2 billion and 2 for China, 4.8 billion and 19.4 for the United States. In 2006 overall Chinese emissions reached American levels. In 2007, they were 5.2 percent above (see Table 8.2). Forecasts by the International Energy Agency (IEA) indicate that the increase in China's CO_2 emissions between now and 2030 – 4 additional gigatons (Gt) – would constitute 40 percent of all additional world emissions, nearly double the new emissions of the older industrialized countries.[19] This situation raises a very difficult ethical question. China's right, as well as the right of all other 'emerging' countries to economic growth and development cannot be put into question. But how can this entirely legitimate right be guaranteed without nullifying international efforts to limit GHG emissions? What is more, the effects caused by the

Table 8.1 Emissions of all anthropogenic gases

Group	1970	2005	2050
OECD	13.7	18.7	23.5
BRIC	5.9	16.1	26.2
ROW	4.3	12.1	21.7
Total baseline	23.9	46.9	71.4

Note: Baseline (figures in $GtCO_2eq$)

Source: OECD (2008), *OECD Environmental Outlook to 2030*, Paris: OECD, p. 25.

Table 8.2 Global and per capita** US and Chinese CO_2 emission between 1980 and 2006*

		1980	1990	2000	2006	2007
United States	Global emission	4,661.6	4,863.3	5,693.0	5,698.3	5,769.3
	Emission per capita	20.5	19.4	20.1	19.0	19.1
China	Global emission	1,420.0	2,244.0	3,077.6	5,645.2	6,071.2
	Emission per capita	1.4	2.0	2.4	4.3	4.6
US emission per capita/Chinese emission per capita		14.6	9.7	8.4	4.4	4.1

Notes:
* Expressed in million metric tons of carbon dioxide.
** Expressed in metric tons of carbon dioxide.

Source: IEA *CO_2 Emissions From Fuel Combustion, Highlights* (2009 Edition).

size of the populations of the major emerging countries make it impossible for us to argue solely in relative terms. The fact that each American emits on average 4.1 times more CO_2 per annum than a Chinese person cannot in any way lead us to conclude that, in the interest of 'justice', no measures for controlling Chinese emissions should be taken so long as Chinese individual emissions lag behind American ones. Indeed, if individual emissions in China reached levels comparable to those in the US, one can easily imagine the deleterious effects on world climate.

Yet another element makes the situation even more complex: the countries which have (on average) the biggest ecological footprint have also the highest HDI (Human Development Index). And those which have the highest HDI have also (always on average) the highest GDP per capita (see Table 8.3).

Similar comparisons can be made between the United States and China (see Table 8.4).

Jean Gadrey has highlighted a strong linear correlation between CO_2 emissions per capita and GDP per capita, up to US$13,000 per capita, when GDP per capita increases by US$3,000 GHG emissions increase by 1 ton. After US$13,000 (only 36 countries in the world out of 177 ranked by the UNDP), this relation weakens.[20] To get a simple idea of the margin of progress that can be achieved by the US, it is useful to have in mind examples such as Japan, which has an HDI of 0.960 and a CO_2 emission per capita of 9.68 tons, or France where these two figures are respectively 0.961 and 5.81 (2007 figures).

Table 8.3 Development indicators and CO_2 emissions

Human Development Index	Number of countries	GDP per capita (PPP US$) 2005	Life expectancy at birth (years) 2005	Combined gross enrolment ratio for primary, secondary and tertiary education (%) 2005	CO_2 emissions (Mt CO_2) 2004	Share of world total* (%) 2004	CO_2 emissions per capita (t CO_2) 2004
High human development $1 \geq HDI \geq 0.8$	70	23,986	76.2	88.4	16,616	57	10.1
Medium human development $0.8 > HDI \geq 0.5$	85	4,876	67.5	65.3	10,215	35	2.5
Low human development $0.5 > HDI \geq 0$	22	1,112	48.5	45.8	162	1	0.3
World	177	9,543	68.1	67.8	28,983	100*	4.5

Note: *The world total includes carbon dioxide emissions not included in national totals. These emissions amount to approximately 5 percent of the world total.

Source: UNDP (2007), *Human Development Report 2007/2008*, pp. 69, 232.

Table 8.4 USA and China: HDI and CO_2 emissions

	HDI	GDP per capita (PPP US$) 2007	Life expectancy at birth (years) 2007	Combined gross enrolment ratio in education (%) 2007	CO_2 emissions (Mt CO_2) 2007	Share of world total (%) 2007	CO_2 emissions per capita (t CO_2) 2007
United States	0.956	45,592	79.1	92.4	5,769.3	19.9	19.1
China	0.772	5,383	72.9	68.7	6,071.2	21.0	4.6

Source: UNDP (2009), *Human Development Report 2009* and IEA, *CO_2 Emissions from Fuel Combustion, Highlights* (2009 Edition).

Notwithstanding the problems discussed above, the target is clear. To avoid systemically dangerous climate change, the rise of the earth's temperature must not exceed 2°C during this century. Above this threshold, the risk of 'large-scale human development setbacks and irreversible ecological catastrophes will increase sharply'.[21] However, if we want to limit the temperature increase to 2°C above the pre-industrial level, the GHG concentration must not exceed 450 ppm (as noted above, this concentration has risen from 280 to 379 over the past 100 years). At 450 ppm the probability to stay beneath 2°C is 50 percent. At 550 ppm this figures shrinks to 20 percent. To fulfill the objective, the carbon budget for the whole twenty-first century must be limited to 1,456 $GtCO_2$, that is to say 14.5 $GtCO_2$ per year.[22] This figure must be compared to present annual emissions: 28.9 $GtCO_2$! (See Table 8.3.)

A number of scenarios have been suggested. The UNDP has identified a pathway to keep the planet under the 2°C threshold. World emissions, after a peak around 2020, would have to fall by 50 percent by 2050 (from a 1990 base-year), and toward zero in net terms by the end of the century. Since the burden of such a mitigation strategy must of course be equitably shared between countries of different development levels, the UNDP distinguishes two groups of counties: industrialized and developing. The first would have to target an emissions peak around 2012, then a 30 percent cut by 2020 and 80 percent by 2050. In the second group there would be large variations. Major emitters would maintain a trajectory of rising emissions to 2020, peaking at around 80 percent above current levels, with cuts of 20 percent as compared to 1990 levels by 2050.[23] These goals are very ambitious and are based on the assumption that there is a universal obligation to avoid negative outcomes. They require imagining 'efficient policies', that is to say policies that permit to reach these objectives (the 2°C threshold needn't be justified here[24]) at the lowest possible costs. This question is however inextricably linked to the question of the burden sharing rules.

BURDEN SHARING RULES

Climate stability being a common resource, collective action is needed to preserve or, in a certain sense, create it. Reduction and adaptation measures are thus required. Reduction (or 'abatement' or 'mitigation') means the curbing of GHG emissions and adaptation means a set of actions designed to prepare humanity to meet the challenges of the rise of the temperature that is, whatever is decided, underway. If the ecological necessity to act is obvious, this requirement is linked to the aim of improving human development. Thus, the new climate regime to be conceived and implemented must place the right to development at its core. Collective choices and commitments must therefore be taken at the world level. The UNFCCC aptly summarizes some of the major challenges that are (still) before us when it underlines that 'the global nature of climate change calls for the widest possible cooperation by all countries and their participation in an effective and appropriate international response, in accordance with their common but differentiated capabilities and their social and economic conditions'. The problem to solve is thus a question of distributive justice concerning the use of carbon sinks or, to put it differently, a problem of sharing the world's carbon budget in a way that will avoid an increase of temperature superior to 2°C in the next 100 years. As James Garvey underlines it: 'The carbon sinks of our world are a finite resource which has been shared out unequally. Justice demands that we redress the balance.'[25]

The 'Relevant Agents' Question

In the climate regime to come it will thus be necessary to define an acceptable regime of burden sharing among economic agents. But, who are the relevant agents? To put it differently: 'Who is obligated to act and to aid?'[26] Traditionally, the answer has been 'the nation-state'. In the debate on burden sharing rules, arguments are frequently based on aggregate national emissions and emissions per capita. All comparisons between states or between individuals take nation-states as the unit and building block of their reasoning. This state-centric approach has a perverse effect since it masks emissions differentials within states among social classes. As Paul Harris argues, many of the solutions to climate change will, of course, have to involve states. 'But this reality needs not absolve capable *individuals* from explicit responsibility and obligation, especially when states are not doing nearly enough.'[27] He rightly asks: 'Why should a poor person in, say, Germany be lumped with the wealthy of Germany to aid both the poor *and* the rich in China or other developing countries who suffer from the effects of climate change, especially when the latter pollute far more?'[28] The question is not simply theoretical, the rising Indian and Chinese middle classes being far more numerous than the German population.

A serious answer must thus address both levels. The state-centered approach must lead to an international climate justice framework with binding national obligations. The social class or citizen-centered approach requires a cosmopolitan or global climate justice framework with individual obligations.

International Climate Justice

The principle of states' obligations founds the international climate justice approach. But how should we define the 'obligations' of any particular state? At least two criteria can be taken into consideration: the historical responsibilities of the state considered and its present (economic, technological, and so on) capacities to contribute to its mitigation. Concerning the first, is it useful to stress that some states have contributed far more than others to the present rises of earth temperatures? The disparities in current levels of GHG emissions reflect the disparities in cumulative emissions since the industrial revolution. The US, for instance, which is the first per capita emitter of the planet and until 2006 was the first emitting country, is responsible for almost 30 percent of cumulative CO_2 emissions between 1850 and 2002. The European Union ranks second with 26.5 percent and Russia third with 8.1 percent. From 1850 till now, the 'developed' world (the historic centre of world capitalism) has been responsible for 76 percent of CO_2 emissions, a figure that means that the responsibility of the 'developing' world (or former peripheries) is limited to 24 percent.[29] These figures clearly show that some countries have made disproportionate use of the carbon sinks of the earth. Second, it is equally obvious that the highly industrialized states have greater financial and technological capacities to develop climate friendly or more energy efficient technologies.

Burden sharing principles must thus be adopted taking the above into account. But, the difficulty arises that there are different kinds of rules relying on different moral justifications. Martino Traxler proposes an interesting typology that distinguishes between 'just' and 'fair' principles. Just principles are 'backward looking' 'historical rectificatory principles' that are 'intended to restore an acceptable moral order that past actions had disturbed'. Examples of 'just' proposals could be: (1) to pay or contribute in proportion to the *benefits* received from the total historical emissions of GHGs; (2) to pay in proportion to the total historical emissions of GHGs; (3) to pay in proportion to responsibility, the responsibility being for instance limited to emissions after a given year of reference, for example 1990. (In fact, in the field of pollution, no one can be held responsible for damage caused at a period when science had not evidenced such damage. That is why the year of the publication of the first IPCC can be considered an acceptable starting point). Fair principles are 'forward looking'. They 'seek to maintain matters at least as morally acceptable as they are found to be at present in the future'. An example of a 'fair' proposal is to pay on an equal per capita basis.[30]

Paul Baer, Tom Athanasiou, Sivan Kartha and Eric Kemp-Benedict

– authors of *The Greenhouse Development Rights Framework* – propose a 'Responsibility Capacity Index' (RCI) to calculate national climate obligations. First, they fix a 'development threshold', that is to say an income level (US$7,500 per capita and per year, in purchasing power parity (PPP)) at which people achieve acceptable Millennium Development Goals indicators. On this basis, a nation's capacity (C) is the sum of all individual incomes above the threshold. Its responsibility (R) is likewise defined as cumulative emissions since 1990, excluding emissions that correspond to consumption below the development threshold. These measures can be combined in the 'Responsibility Capacity Index'.[31] With a $RCI = aR + bC$ (with $a = b = 0.5$) we obtain the following figures (see Table 8.5).

Because the measure of capacity excludes the income of poor people, a rich country's capacity will be larger in percentage terms than its share of global income, and a poor country's capacity will be smaller. Likewise, a wealthy country's responsibility will be larger than its share of cumulative emissions because fewer of its historical emissions will be excluded. If we assume that the total cost of the global climate program is 1 percent of gross world product, or about US$1 trillion in 2020, we obtain the following estimates of obligation to pay (for mitigation and adaptation) (see Table 8.6).[32] (If the cost is 2 percent, the last two columns have to be multiplied by 2.)

Even if these figures seem huge, they must be considered in a perspective of economic growth. As Christian Azar and Stephen Schneider demonstrate, the abatement cost of global warming would be overtaken after a few years of income growth:

> If the cost by the year 2100 is as high as 6% of global GDP and income growth is 2% per year, then the delay time is 3 years, whereas the delay time is only 1 year if income grows by 3% per year and the abatement cost is 3% of GDP.[33]

If you rank countries according their average obligation per person above the development threshold, it appears that 17 of the top 40 countries are not countries included in the Annex 1 of the Kyoto protocol.[34] But states are not the only relevant actors obligated to act. What about affluent people, whatever their nationality? Taking this question into consideration leads to the concept of cosmopolitan climate justice.

Cosmopolitan Climate Justice

As Paul Harris argues, the cosmopolitan conception of justice defines individuals as world citizens: 'A cosmopolitan approach places rights

Table 8.5 *Greenhouse Development Rights Framework for the United States and China*

	2010					2020	2030
	Population (% of global)	GDP per capita ($ US PPP)	Capacity (% of global)	Responsibility (% of global)	RCI (% of global)	RCI (% of global)	RCI (% of global)
US	4.5	45,640	29.7	36.4	33.1	29.1	25.5
China	19.7	5,899	5.8	5.2	5.5	10.4	15.2
World	100.0	9,929	100.0	100.0	100.0	100.0	100.0

Source: Baer et al. (2008), *The Greenhouse Development Right Framework*, Heinrich Böll Foundation, Christian Aid, EcoEquity and the Stockholm Environment Institute, p. 55.

Table 8.6 National obligations to pay

	National income (billion US$)	National capacity (billion US$)	National capacity (% GDP)	National obligation (billion US$)	National obligation (% GDP)
United States	18,177	15,661	86.2	275	1.51
China	13,439	5,932	44.1	98	0.73
World	94,405	59,388	62.9	944	1.00

Source: Baer et al. (2008), *The Greenhouse Development Right Framework*, Heinrich Böll Foundation, Christian Aid, EcoEquity and the Stockholm Environment Institute, p. 58.

and obligations at the individual level, discounting the importance of national boundaries.'[35] With this perspective, national boundaries are not considered a morally distinctive feature for the elaboration of burden sharing rules. The cosmopolitan conception of justice can thus be seen as an application, at the international level, of John Rawls' veil of ignorance. As Charles Beitz puts it: 'For purpose of moral choice, we must . . . regard the world from the perspective of an original position from which matters of national citizenship are excluded by an extended veil of ignorance.'[36] But, why should such a shift from international to cosmopolitan justice be considered so important? The answer is quite simple: in 'emerging' countries a growing number of people enjoy middle-class lifestyles in terms of consumption patterns, and the number of these 'new consumers'[37] is going to increase substantially in the years to come.

Homi Kharas and Geoffrey Gertz define these new middle classes as those households with daily expenditures between US$10 and US$100 (PPP), that is to say between the average poverty line in Portugal and Italy and twice the median income in Luxemburg. In a text published in 2010, they forecast that in 2022 the world middle class population will be larger than the poor population and that, by 2030, 5 billion people, or nearly two-thirds of the world population, could belong to the middle class.

> According to our estimate, by 2015, for the first time in 300 years, the number of Asian middle class consumers will equal the number in Europe and in America. By 2021, on present trends, there could be more than 2 billion Asians in middle class households. In China alone, there could be over 670 million middle class consumers, compared with only perhaps 150 million today.[38]

According to the authors Chinese middle class consumption totaled US$859 billion in 2009 (in 2005 PPPUS$). The Chinese car fleet rose from 1.6 million in 1990 (152 million in the United States) to 8 million in 2000

(175 million in the United States, this figure including SUVs), and reached 53 million in 2007.[39] The growing importance of this emerging middle class is evidenced when GNP (gross national product) or GDP (gross domestic product) are expressed (and compared) not in international-exchange dollars but in PPP. Using PPP provides an indicator of wellbeing that is free of exchange rates distortions. This leads to sometimes substantial modifications in country rankings (see Table 8.7).

PARTIAL COSMOPOLITANISM

Because of its global scope, the first crisis of its kind in recorded history, global climate change poses a major challenge for political philosophy in general and international relations theory in particular. It follows from what has been said that the normatively best approach would be based on a holistic assumption of species and eco-systemic unity, and of universal interdependence transcending national and cultural distinctions. This in turn implies the inalienable right of all individuals, present and future, to a 'good life', that is a life worth living 'with and for others under just institutions'.[40] In a cosmopolitan setting, the distributional issues raised by the need to offset global climate change would be considered a problem of social justice among individuals and social classes at global level rather than between nation-states. This approach presumes the need for empowered institutions of global governance, the purpose of which would be to define and implement universal public policies designed to secure inter and intra generational equity. In the pursuit of this set of normative goals, theory should be geared towards generating a cosmopolitan ethos and shared inter-subjective meanings regarding the human prospect based on solidarity.

The difficulties of implementation of such a cosmopolitan perspective are however enormous. It presupposes the passage from the modern Westphalian international system, which is segmented into

Table 8.7 Two of the world's largest economies in 2008 (GDP)

	US$, billions	Rank	PPPUS$, billions	Rank
United States	14,093	1	14,093	1
China	4,327	3	7,909	2

Source: The Economist, *Pocket World in Figures. 2011 Edition*, London: The Economist and Profile Books Ltd, 2010, p. 24.

sovereign national territorial units, to a post-modern configuration of post-nationality. Yet the imagined communities we live in and have constructed since the rise of the modern nation state are based on ontological assumptions about identity, belonging and obligations that cannot be simply swept away by new scientific knowledge (this is another way of saying what we affirmed earlier regarding descriptive and normative judgments).[41] National, religious, ethnic and racial segmentation remain stubborn if unfortunate social facts. Despite deepened transnational linkages at multiple levels, the world system is still very far from the post-international and post-national politics envisioned by some political theorists in the aftermath of the Cold War.[42] For reasons relating to past experiences of subordination, post-colonial states tend to adhere to modern criteria of power and sovereignty. Notwithstanding the internationalization of business and other elite segments, so does the United States. Even in the European Union, national segmentation remains a hindrance to transnational governance and accomplished federalism.

Even if it seems that a shared understanding is gradually emerging that the collective human fate is inescapably bound to finding global answers to transnational problems, the abstract understanding of shared humanity (species-being, in Marx's formulation[43]) does not automatically translate into a cosmopolitan ethos since it runs counter to the daily experience of difference, otherness and self-interest (however flawed and ultimately self-defeating these can be shown to be). Likewise, even if we are able to rid ourselves, individually and collectively, of philosophies of radical selfishness (*homo economicus*) this does not mean that they will spontaneously give way to empathy towards 'strangers', much less universal altruism. While the statement 'we are all likely to die or at least to suffer extremely deleterious consequences if we don't share the burden of global climate change' is certainly accurate, it does not tell us how in fact various nations and social classes should share it, or who will define the terms of the settlement. Lastly, even assuming that global institutions can be created to find ways out of our present collective predicament, how to ensure that they will be just?

Ordered Pluralist Cooperation

These are extraordinarily difficult problems. To avoid insuperable contradictions we should start by rejecting options that are either undesirable or unattainable. The first undesirable outcome of the eco-systemic crisis would be a sharpening of international segmentation through an exacerbation of competitive struggles over scare resources, an outcome predicted by realist and neo-realist international relations theory. Still dominant in the contemporary literature, neo-realism postulates that nation-states are functionally undifferentiated self-seeking units that are conditioned by the anarchic structure of the international system to maximize their power and minimize their insecurity, to the exclusion of all else. Under conditions of eco-systemic crisis, this Hobbesian assumption, which has been subjected to sharp epistemic and methodological critique,[44] implies an inevitable sharpening of interstate conflict. Hence, neo-realism has to be excluded outright as a relevant framework in dealing with global climate change. However, as Richard Falk has rightly pointed out, we also have to exclude a number of 'post-Westphalian scenarios' as either undesirable or unattainable or both, among them the neo-liberal utopia of a self-regulating global market system, the limits of which are now plainly apparent, or the idea of a world government.[45] If it were possible, which it is not, world government would erase pluralism and is not synonymous with democratic global governance. A world Leviathan would more likely reproduce at global level, under highly coercive circumstances, the social inequalities that presently exist within national boundaries, than establish global justice. Michael Walzer makes a similar point in his typology of various international orders,[46] which establishes a spectrum according to their degree of centralization. He cogently argues that while 'the decentered world' of international anarchy threatens peace, the 'tyrannical potential' of a unified global state or a global empire would undermine individual liberty and cultural pluralism. The problem thus is to 'overcome the radical decentralization of sovereign states without creating a single all-powerful central regime'.[47]

The alternative would be a 'global pluralism', or a pluralistic international society, underpinned by reinforced and empowered institutions of global governance. Falk points to an evolutionary solution when he calls for the 'gradual emergence of an accountable global polity'.[48] To be legitimate, that global polity would be inclusive, democratic, pluralistic and founded on human solidarity. Imagining that polity is a core theoretic challenge posed by global climate change. A first step in that direction would be to go back to and to renew the basic principles enunciated in the United Nations Charter regarding universal social, economic and human rights, with the dual aim of achieving greater fairness among nations as well as among individuals and social classes at global level. Both require democratizing the decision-making and reshaping of the normative orientation of the multilateral institutions responsible for world economic governance (IMF, World Bank, WTO, and so on). Increasing the voting rights of 'emerging' as well as poor countries would enhance democratic fairness among nations. At the same time, the composition of the UN Security Council must be made to reflect world diversity and plural interests. *A minima* this implies ending the monopoly of the five permanent

Table 8.8 Possible political arrangements of international society

← Centralization/unity				Decentralization/division →		
Unified global state (1)	Global empire (2)	Federation of nation-states (3)	Third degree of global pluralism (4)	Second degree of global pluralism: 'Strong' international society (5)	First degree of global pluralism: 'Weak' international society (6)	International anarchy (7)

members by giving decision-making positions to countries such as Brazil, India, Japan, Germany and South Africa. Going a bit further one could imagine a system of representation at regional or sub-regional level (European Union, Sub-Saharan Africa, South America, and so on).

While necessary, greater fairness among nations and more democratic interstate relations are hardly sufficient to move towards cosmopolitan justice. It would have to be accompanied by greater social fairness worldwide to secure the basic economic and social rights of individuals. The renovated institutional system would thus have as mission to act at global level to reduce poverty, secure food and water supply and define common humane norms for dealing with migrations and other pressing issues of human survival. As Pierre Bauchet has suggested, these aims could be constitutionalized through a redefinition of the tasks of the international public institutions, their relations and the principles of their interventions.[49] These include: the principle of subsidiarity, the principle of global regulation (harmonization of fiscal regimes, employment conditions) and the principle of consensus-based management. In a similar vein, Mireille Delmas-Marty has suggested that the alternative to the chaos of interstate rivalry or to hegemonic rule by a single world state or a world empire would be incremental movement towards 'ordered pluralist cooperation'.[50] The concept implies the gradual convergence of actor agendas around common goals across different issue areas. Rather than erasing pluralism, it would seek to identify areas of convergence that allow for cooperative action in a pluralistic setting, successful cooperation in any one area generating trust and opening the possibility for advances in other areas. In contrast to hegemonic regimes, the rules and disciplines of which are set by a dominant power, ordered pluralism would not be based on hierarchy but on the mutual needs of various actors with different capabilities.

What is being suggested here is to eschew attempts to achieve an unattainable *complete* cosmopolitan order in favor of a partial cosmopolitanism that does not erase plural traditions and identities[51] and that would allow movement towards convergence and deepened interdependence. These concepts can be usefully applied to the question of global warming by identifying areas of congruence between national and universal interests.

Convergence Around Norms

A good concrete example of possible cooperation and convergence around common goals is Nicholas Stern's 2009 proposal of a 'global deal'.[52] 'Global' both in its origins and in its impacts that deal, he writes, must be *effective* (in the sense that it reduces GHG emissions on the scale required), *efficient* (for example at the lowest possible costs) and *equitable* (according to the responsibilities and abilities of the different actors involved). Stern's deal must be understood as an 'integrated package' containing six elements that can be brought together in two categories: 'targets and trade' and 'funding'.[53] Concerning the 'targets and trade', Stern defines three goals: the world GHG emissions must be at least reduced by 50 percent by 2050 relative to 1990; developing countries should commit themselves to take on targets at the latest by 2020 if developed countries have fulfilled their own commitments. Country emissions reductions and carbon trading schemes must be adopted and designed in order to integrate trading mechanisms with other countries, in particular with developing countries. The 'funding' should be assured by three series of measures that can be summarized as follows: strong initiatives, with public funding, to build capacity to stop deforestation; development, demonstration and sharing of technologies; commitments on overseas development based on Monterrey (2002 UN), UE 2005 and Gleneagles (2005 G8) – rich countries must deliver assistance in the context of extra costs of development arising from climate change.

Another promising proposal for a fair global settlement was made in 2009 by Hu Angang, a senior figure of the Chinese Academy of Sciences. It would take into account 'average greenhouse-gas emissions per capita, total greenhouse-gas emissions, [and] historical and current responsibilities.'[54] In Hu's proposal, the Human Development Index (HDI) of countries would serve as the metric of burden sharing rather than GDP. Thus, various categories of countries – High (above 0.8), Medium High (between 0.65 and 0.8), Medium Low (between 0.5 and 0.65) and Low HDI (under 0.5) – would contribute differentially to the reduction of emissions. If all the largest polluters, which include both High HDI (the 'developed countries') and Medium High HDI countries (such as China, HDI = 0.772 in 2007), will be required to make major efforts to cut back emissions, the High HDI countries, which owe their present position of wealth and power to their historical position at the core of the hierarchical late modern world system, would be called upon to make the greatest unconditional efforts not only in terms of emissions reduction but also in terms of providing financial and/or technological assistance to the other groups of countries. This would help to significantly mitigate climate warming while correcting the patterns of international inequality set in the nineteenth century and that still shape the world today. Medium Low HDI countries would 'benefit from low-interest loans from international financial organizations and low-cost technological assistance'. Obligations of various countries would of course evolve according to their shifting position in the Human Development Index. Thus Medium High HDI countries would become unconditional reducers once they have reached High HDI status. The global effort would be enforced by a United Nations agency established to that effect and which would set binding targets for all countries.

Global distributive justice would have to be complemented by fair social bargains at national level. At the national level distributive justice in burden sharing would be accomplished through redistributive policies such as progressive tax policies, incentives to invest in green technologies, punishing disincentives for polluters, and so on. In other words, the global bargain would have to be sustained by domestic social democratic bargains based on a fair distribution of resource consumption among social classes and a reorientation of patterns of consumption.

CONCLUSION

Current world leaders have an immense responsibility, since their actions or non-actions will determine whether survival oriented outcomes prevail or not. At this time of writing, the preparatory negotiations prior to the November–December 2010 Cancun Summit remain bogged down in acrimonious disputes between 'emerging' and highly industrialized states regarding historic responsibilities for global climate change. As far as China and the US are concerned, both countries' agendas on climate change are contaminated by competition in other issue areas (currency rates, rival security claims in the South China Sea, and so on). The world economic crisis that broke out in 2007 has accentuated competitive pressures and the temptation of states to engage in 'beggar their neighbor' trade and currency policies.

The obstacles to a fair global settlement are thus very great. But failure is not an acceptable outcome. Change from on top will require pressure from below.[55] It also requires a concerted conceptual effort to overcome the 'theoretical ineptitude' discussed at the outset, to generate the intellectual and hence the political conditions to initiate movement towards convergence around shared understandings and norms. In the social sciences, theory is 'reflexive' in the sense that the knowing subject emits judgments and interpretations that modify the object of knowledge, leading to the perpetual construction and re-construction of 'reality'. Theorists have the responsibility to construct a cosmopolitan ethos that will underpin a world order based on pluralist cooperation rather than rivalry.

NOTES

1. Martin Rees, 'Science: The Coming Century', *The New York Review of Books* 55, no. 18 (20 November 2008): 41.
2. Stephen M. Gardiner, 'A Perfect Moral Storm: Climate Change, Intergenerational Ethics and the Problem of Moral Corruption', *Environmental Values* 15 (2006): 397.
3. David Archer, 'Fate of Fossil Fuel CO_2 in Geologic Time', *Journal of Geophysical Research* 110 (2005): 5, doi:10.1029/2004JC002625.
4. Gardiner (2006), p. 407.
5. OECD (Organisation for Economic Co-operation and Development), *OECD Environmental Outlook to 2030* (Paris: OECD, 2008), pp. 141–143.
6. To borrow the expression of Article 2 of the UN Framework Convention on Climate Change of 1992 (UNFCCC).
7. William Nordhaus, *A Question of Balance. Weighing the Options on Global Warming Policies* (New Haven & London: Yale University Press, 2008), p. 62.
8. James Garvey, *The Ethics of Climate Change. Right and Wrong in a Warming World* (New York: Continuum, 2008), p. 61.
9. Joseph Stiglitz, 'Knowledge as a Global Public Good', in Inge Kaul, Isabelle Grunberg and Marc Stern (eds) *Global Public Goods. International Cooperation in the XXIst Century* (New York: Oxford University Press, 1999), pp. 308–325.
10. Will Steffen, Paul J. Crutzen and John R. McNeill, 'The Antropocene: Are Humans Now Overwhelming the Great Forces of Nature?', *Ambio* 36, no. 8 (December 2007): 614–621.
11. Martino Traxler, 'Fair Chore Division for Climate Change', *Social Theory and Practice* 28, no. 1 (January 2002): 120.
12. Paul Samuelson and William D. Nordhaus, *Economics* (New York: McGraw Hill, 1985), p. 713.
13. Garrett Hardin, 'Tragedy of the Commons', *Science* 162, no. 3859 (13 December 1968): 1243–1248.
14. Gardiner (2006), p. 400.
15. Joseph Stiglitz, 'A New Agenda for Global Warming', *The Economists' Voice*, 3, no. 7 (July 2006), article 3, p. 2; available at http://www.bepress.com/ev/vol3/iss7/art3.
16. Robert Axelrod, *The Evolution of Cooperation* (New York: Basic Books, 1984).
17. Gardiner (2006), p. 404.
18. *Ibid.*, p. 405.
19. IEA (International Energy Agency), *World Energy Outlook 2006* (Paris: OCDE/AIE, 2006), p. 188.
20. Jean Gadrey, 'Croissance, bien-être et développement durable', *Alternatives économiques* 266 (February 2008): 70.
21. UNDP (United Nations Development Programme), *Human Development Report 2007/2008* (New York: UNDP, 2007), p. 7; available at www.undp.org.
22. *Ibid.*, p. 46.
23. *Ibid.*, p. 48.

24. As it seems to exist a consensus (or at least a 'near-consensus') on this figure and on the pathway of GHG emissions reduction the world has to follow, we will not discuss here what William Nordhaus for instance call the 'when efficiency requirement', that is to say the level of the rate of social time preference that must be applied to the cost–benefit analysis (Nordhaus, 2008). The likelihood of a 'dangerous' climate change if the rise of temperature should exceed 2°C seems great enough to justify action.
25. Garvey (2008), p. 76.
26. Paul Harris, 'Climate Change and Global Citizenship', *Law & Policy* 30, no. 4 (October 2008): 482.
27. *Ibid.*, p. 483.
28. *Ibid.*, p. 484.
29. Garvey (2008), p. 70.
30. It is not useless to note that in 2004 the CO_2 world emissions per capita was 4.5 tons, a figure to compare to the Chinese figure: 3.8 tons and to the American one: 20.6 tons. These figures are obtained with a world total emission of almost 29 gigatons. With the annual budget of 14.5 gigatons cited above, these figures should (with the same population) be reduced by 50 percent! We understand why Chinese authorities prefer 'intensity targets' (carbon intensity of economic growth, energy intensity of the economy) to 'absolute targets'. For a critique of a contribution on equal per capita basis, see Traxler (2002), pp. 124–125.
31. Paul Baer, Tom Athanasiou, Sivan Kartha and Eric Kemp-Benedict, *The Greenhouse Development Right Framework. The Right to Development in a Climate Constrained World* (Heinrich Böll Foundation, Christian Aid, EcoEquity and the Stockholm Environment Institute, 2008).
32. Explanation: the figure US$275 billion for the United States (column 5) is obtained the following way: 944.05 x 0.291 (from Table 8.5 column 7) = 274.7 (944.05 is 1 percent of world GDP).
33. Christan Azar and Stephen H. Schneider, 'Are the Economic Costs of Stabilising the Atmosphere Prohibitive?,' *Ecological Economics* 42 (2002): 77.
34. Baer et al. (2008), p. 62.
35. Harris (2008), p. 486.
36. Charles R. Beitz, *Political Theory and International Relations* (Princeton, NJ: Princeton University Press, 1979), p. 176.
37. Norman Myers and Jennifer Kent, *The New Consumers. The Influence of Affluence on the Environment* (Washington, Covelo, London: Island Press, 2004).
38. Homi Kharas and Geoffrey Gertz, 'The New Global Middle Class: A Cross-Over from West to East' (Wolfensohn Center for Development at Brookings, 2010), p. 2; available at http://www.brookings.edu/~/media/Files/rc/papers/2010/03_china_middle_class_kharas/03_china_middle_class_kharas.pdf (accessed 4 February 2011).
39. For a detailed presentation of the situation, see Jean-François Huchet and Jean-Paul Maréchal, 'Ethics and Development Model', *China Perspectives* 1 (2007): 6–16.
40. Paul Ricœur, *Soi-même comme un autre* (Paris: Seuil, 1990), p. 202.
41. Benedict Anderson, *Imagined Communities: Reflections on the Origins of Nationalism* (London: Verso, 1991).
42. James Rosenau, Ernst-Otto Czempiel, *Governance without Government: Order and Change in World Politics* (Cambridge: Cambridge University Press, 1992); Jürgen Habermas, *La paix perpétuelle, le bicentenaire d'une idée kantienne* (Paris: Le Cerf, 2006).
43. Global warming confronts individuals to the existential reality of species-being, in the sense given to that concept by Marx. See Karl Marx and Friedrich Engels, *Collected Works*, Vol. 3 (New York: International Publishers, 1973).
44. See for instance Richard Ashley's and Robert Cox's critiques in *Neorealism and its Critics*, Robert Keohane (ed.) (New York: Columbia University Press, 1986).
45. Richard Falk, *The Declining World Order: America's Imperial Geopolitics* (London: Routledge, 2004).

46. Michael Walzer, 'Governing the Globe: What Is the Best We Can Do?', *Dissent Magazine* (Fall 2000); available at http://www.dissentmagazine.org/article/?article=1436.
47. *Ibid.*
48. Falk (2004), p. 9.
49. Pierre Bauchet, *Concentration des multinationales et mutation des pouvoirs de l'État* (Paris: CNRS Editions, 2003).
50. Mireille Delmas-Marty, *Les forces imaginantes du droit : Tome 2, Le pluralisme ordonné* (Paris: Seuil, 2006).
51. Kwame Anthony Appiah, *Cosmopolitanism: Ethics in a World of Strangers (Issues of Our Time)* (New York: W.W. Norton, 2007).
52. Nicholas Stern, *The Global Deal. Climate Change and the Creation of a New Era of Progress and Prosperity* (New York: Public Affairs, 2009), p. 3.
53. *Ibid.*, pp. 146–147.
54. Angang Hu, 'A New Approach at Copenhagen (1) (2) (3)', *Chinadialogue* (6 April 2009), available at http://www.chinadialogue.net.
55. Jessica Seddon Wallack and Veerabhadran Ramanathan, 'The Other Climate Changers. Why Black Carbon and Ozone Also Matter', *Foreign Affairs* 88, no. 5 (September/October 2009): 105–113.

9. Cosmopolitan diplomacy and the climate change regime: moving beyond international doctrine

Paul G. Harris

INTRODUCTION

Efforts by governments and the international community over the last three decades to cooperate in protecting the global environment have failed to bring about robust action to limit greenhouse gas pollution causing climate change. While pursuing apparently logical economic and social development, and by acting in ways that seem to be promoting the interests of nation-states and their citizens, humanity continues to dangerously alter the Earth's atmospheric and climate systems, with profound consequences for human well-being and, for many millions of people, even survival. One reason for this tragedy is the preoccupation of governments and societies with political independence and national sovereignty, the existence of an international system premised on that sovereignty and a failure to adequately recognize twenty-first century realities, notably rapidly expanding numbers of new consumers in the developing world that are adding greatly to the greenhouse gas pollution that has long come from the developed world. The dilemma brought on by a preoccupation with states and their sovereign rights requires an alternative pathway leading to environmentally sustainable development that is agreeable to both rich and poor countries and peoples.

This chapter proposes a way forward for the climate change regime and resulting policies that acknowledges the responsibilities and duties of developed states while also explicitly acknowledging and acting upon the responsibilities of all affluent people, regardless of nationality.[1] My aim is to explore the role of justice in the world's policy responses to climate change, and in particular to describe an alternative strategy for tackling climate change that is more principled and practical than the prevailing approach, and which may be much more politically acceptable to governments and citizens than are existing responses to the problem. This alternative strategy is premised on cosmopolitanism. A cosmopolitan ethic, and its practical implementation in the form of global justice, offers both governments and people a path to sustainability and successful mitigation of the adverse impacts of climate change.

My argument in favour of a more cosmopolitan approach to dealing with climate change is not meant to be an idealistic exercise or an act of imploring the world to come around to the view that all people will soon feel that they are global citizens or that states can be abandoned. Rather, this is an attempt to show that the most practical and politically viable approach to climate change – as well as the most principled one – is in fact one that actualizes cosmopolitan ethics, and more often than not can be and should be premised upon those same ethics. The upshot is that by placing persons – and their rights, needs and duties – at the centre of climate diplomacy and discourse, more just, effective and politically viable (and palatable) policies are more likely to be formulated and implemented.

FROM INTERNATIONAL TO GLOBAL JUSTICE

Global warming is causing increasingly significant ongoing climate change that will become profoundly damaging to human well-being in this century and beyond. While all regions of the world will be impacted by climate change, it is the poorest regions and poorest people that will suffer the most. The world‘s wealthy countries and people will, in most cases, be able to adapt to climate change, or at least survive it. In contrast, the poorest countries, the poorest regions within them and the world‘s poorest individuals, most of them in Africa and developing parts of Asia and Latin America, will suffer and often die as a consequence of climate change. Historically it has been the world's wealthy states and their citizens that have polluted the atmosphere, often as a result of conspicuous consumption and other activities that are not essential to life or happiness (and indeed often undermine them, as when people neglect family and friends to garner wealth and possessions or when they consume foods that are both bad for the environment and bad for their health). Now the burgeoning middle and wealthy classes of the developing world – the world's hundreds of millions of new consumers – are adding to this pollution, leading to explosive growth in greenhouse gas emissions. Importantly, those who will suffer the most from climate change – the world's poor – are the least responsible for it. This makes climate change a profound matter of justice – and injustice.

International Justice

As governments have sought to address transboundary environmental problems they have layered the concept of environmental justice with longstanding interstate doctrine. Since the Treaty of Westphalia in 1648, the world has been guided by, and governments have sought to reinforce, international norms of state recognition, sovereignty and non-intervention. According to these prevailing and powerful norms, states are the ultimate and most legitimate expressions of human organization, the venues for morality and the solutions to major challenges that extend beyond individual communities. These norms have so far largely guided discourse, thinking and responses to transboundary environmental problems: international environmental diplomacy, regimes and treaties have been based (almost by definition) on the responsibilities, obligations and capabilities of *states* to limit their pollution or use of resources, and to work together to cope with the effects of environmental harm and resource exploitation. The Westphalian international norms have been so powerful as to result in a doctrine of *international* environmental justice, manifested in the principle of common but differentiated responsibility among states. This doctrine has guided the creation of many recent international environmental agreements, but states have been noteworthy for the degree to which they have failed to implement it. This is a consequence of the doctrine itself, which mitigates against cooperation for the common good, instead encouraging promotion of narrow and short-term perceived interests of states. In the case of climate change, Westphalian norms have stifled diplomacy and prevented policy innovations, fundamentally ignoring the rights, responsibilities and duties of *individuals*.

The international climate change regime and its provisions for international environmental justice are premised on interstate doctrine. The doctrine of international environmental justice that has emanated from Westphalian norms, discourse and thinking has taken the world politics of climate change in a direction that has been characterized by diplomatic delay, minimal (or no) action – especially relative to the scale of the problem – and mutual blame between rich and poor states resulting in a 'you go first' mentality even as global greenhouse gas emissions skyrocket. The doctrine is one premised on national interests, which in practice routinely translates into national selfishness. The international doctrine has been written into international agreements such as the United Nations Framework Convention on Climate Change, the Kyoto Protocol and subsequent agreements and diplomatic negotiations on implementing the protocol and devising its successor. Although some major industrialized states in Europe have started to restrict and even reduce their emissions of greenhouse gases, these responses pale in comparison to the major cuts – exceeding 80 percent, *at minimum* – demanded by scientists.[2] Indeed, global emissions of greenhouse gases are *increasing*, and will do so for decades unless drastic action is taken very soon. This is in large part due to huge emissions increases being experienced in major developing countries as their economies grow and as millions of their citizens adopt Western consumption patterns. At present, however, emissions from the expanding wealthy classes and new consumers of the world are excluded from the climate change regime because most of the states in which those people live are victims of pollution from traditional consumers in the world's wealthy countries. This exclusion obtains despite the growing impact of this new consumption and pollution on the Earth's atmosphere.

Global Justice

One potentially potent remedy to the Westphalian norms that plague today's responses to climate change can be found in cosmopolitan ethics and global conceptions of justice that routinely and explicitly consider people as well as states. A cosmopolitan approach places rights and obligations at the individual level and discounts the importance of national identities and state boundaries. Cosmopolitans recognize the obligations and duties of responsible and capable individuals regardless of their nationality. From a cosmopolitan perspective, what matters are (for example) affluent Americans and affluent Chinese *people*, rather than the 'United States' or 'China' qua *states*. People in one state do not matter more than people in others. Generally speaking, *international* justice views national borders as being the basis for justice. In contrast, *global* justice, while accepting that national borders have great importance in the world, sees them as being the wrong basis for justice. This is especially so in the case of climate change.

Perhaps the most important development in the world today is the rise of hundreds of millions of new consumers in a number of developing countries.[3] Barely more than a decade ago it was possible to talk about climate change, both in practical and moral terms, by exclusively pointing to the responsibility of developed countries and their citizens as the causes of atmospheric pollution and as the bearers of the duty to end that pollution, make amends for it and aid those who will suffer from it. The climate change regime, insofar as it recognizes this responsibility, is premised on this notion. But the world has changed dramatically in recent years. The developing countries together now produce fully half of the world's greenhouse gases. China has overtaken the United States to become the largest national source of these pollutants. Given the developing countries' large populations, this change does not in itself alter the moral calculus very much because their national per capita emissions remain low relative to the developed countries. What has changed, however, is the increasing number of new consumers in these countries, many of them very affluent indeed, who are living lifestyles analogous to, and often superior to (in terms of material consumption), most people in the developed countries. Now numbering in the hundreds of millions, these people are producing greenhouse gases through voluntary consumption at a pace and scale never experienced. While many societies in the West are finally starting to make changes that limit and reduce their greenhouse gas emissions, the new consumers are going in the opposite direction, with truly monumental adverse consequences for the atmospheric commons. At present, these new consumers face no legal obligations to mitigate the harm they do to the environment, and they have so far escaped moral scrutiny. If solutions to climate change are to be found, this will have to change, not least because 'old consumers' in developed societies will be watching these new consumers do the things that the old consumers are being told they must not do in order to help the world tackle climate change. As long as the new consumers hide behind their states' poverty, practical and politically viable solutions to climate change will be very difficult to realize.

The remainder of this chapter proposes an alternative to the status quo climate change regime – a regime that is premised on the rights and duties of states while largely ignoring the rights and duties of too many people. Because cosmopolitanism is concerned with individuals, it can help the world reverse the failed national and international policies that have contributed to the tragedy of the atmospheric commons. It can do this in part by addressing the lack of legal obligations for many millions of affluent people in developing countries to limit their greenhouse gas emissions in any way while still recognizing that the world's affluent states, and indeed the affluent people within them, have even more responsibility to do so. Cosmopolitan justice points us to a fundamental conclusion: that affluent people *everywhere* should limit, and more often than not cut, their atmospheric pollution, regardless of where they live. Cosmopolitan aims should be incorporated as *objectives* of climate change diplomacy and policy. This points to a cosmopolitan corollary to the doctrine of interstate justice, one that explicitly acknowledges and acts upon the duties of all affluent people, regardless of nationality, to be good global citizens. A cosmopolitan corollary would comprise a new form of international discourse, a new set of assumptions about what states and their citizens should be aiming for, and a new kind of institutionalism that folds world ethics and global justice into the practice of states. This corollary would be more principled, more practical and indeed more politically viable than current doctrine and norms of international environmental justice applied to climate change.

A COSMOPOLITAN MOVE BEYOND INTERNATIONAL DOCTRINE

If any issue demands a cosmopolitan response, climate change is it. Cosmopolitan justice addresses the disconnect between the lack of any legal obligation of many millions of affluent people beyond the scope of the climate change agreements – including the affluent in developing countries – to cut their greenhouse gas pollution, and their ethical responsibility to cut pollution alongside affluent people living in the few rich states that have agreed to binding national obligations in the context of the climate change regime. Implementing cosmopolitan justice here means that obligations to act on climate change, and to aid people harmed by it, apply to all affluent individuals *regardless of where they live*. This points to a corollary (or supplement) to prevailing applications of international justice to climate change: a way forward for climate justice that acknowledges the responsibilities and duties of developed states while also explicitly acknowledging and acting upon the responsibilities of all affluent people, regardless of nationality, as global citizens. A corollary is more principled, more practical and indeed more politically viable than the current doctrine and norms of international environmental justice applied to climate change.

Here I query some of the most common cosmopolitan arguments about global justice and describe an alternative to the purely international response to climate change, in the process proposing that cosmopolitan aims be incorporated as *objectives* of climate change diplomacy and policy. I describe some of the skeletal features of a cosmopolitan corollary to state-centric policy responses to climate change. I suggest some of the ways by which we might reconcile the failure of the climate change regime to recognize the rights and duties of persons with the pressing need to bring all capable people, and especially affluent people – including the world's new consumers – into the picture. A cosmopolitan corollary to international doctrine recognizes everyone's practical and ethical importance. An objective here is to start outlining how an alternative to international doctrine might be implemented. I do not claim that these are the only or even the best ways of actualizing cosmopolitan justice in this context.

The burden of determining how to implement climate diplomacy must be preceded by an appreciation of the importance of new thinking, discourse and policy that is highly sensitive to cosmopolitan objectives. By associating the wealth and behaviours – and the pollution – of individuals and classes of people with ethical diplomatic arguments, international agreements and the domestic implementation of those agreements, governments of both developed and developing states can escape the ongoing blame-game in which poor countries blame rich ones for the problem so far, and rich states blame poor ones for the problem to come – with both refusing to sufficiently obligate even their affluent citizens to do all that is necessary and just. A new kind of climate diplomacy premised on cosmopolitanism allows major developing country governments to simultaneously assert and defend their well-justified arguments rejecting *national* climate change-related obligations while also acknowledging and regulating growing wealth and pollution among a significant segment of their populations – the affluent. Because of its universal application – because people in developed countries will see new consumers in the developing world taking action – a cosmopolitan corollary in turn can help to neutralize the understandable (if unjust) political reticence of most developed country governments to live up to their states', and their own affluent citizens', obligations to finally undertake the major cuts in greenhouse gas emissions that the Earth requires. Applied to climate change, cosmopolitan justice has the potential to define a pathway whereby *all* countries, both developed and developing, can participate fully in the climate change regime *without* making any demands on the world's poor – indeed, while aiding them in new ways.

Institutionalizing Cosmopolitanism Among States

One might posit that, *especially in the context of climate change*, cosmopolitanism is more realistic than communitarian state-centric approaches to solving global problems. Because cosmopolitanism is premised on the rights and interests of persons, it reveals the true locus of pollution causing climate change and the profound consequences of this pollution for billions of people. However, the way that cosmopolitanism is normally applied to the problem usually falls short because it focuses on only about half of the people causing it – affluent people in developed states – while ignoring affluent people in developing states who are rapidly catching up to their developed-country counterparts in their power to consume and pollute. Furthermore, most cosmopolitan responses routinely revert to state-centric prescriptions, with the usual argument being something like this: affluent people in the developed countries pollute so much, therefore their *states* must reduce national greenhouse gas emissions and aid developing *states*, and possibly compensate them as well. How can we adapt these state-oriented conceptions and practices of climate justice, as well as most of the cosmopolitan alternatives that tend to surrender to state domination as well? Doing so is not at all easy, which is one reason why many cosmopolitans argue the way that they do. Even when they desire institutions and policies based on cosmopolitan morality, they are realistic and recognize that states dominate the world and, for very practical reasons, have to be major actors in finding solutions to climate change. From this thinking it is then normally concluded that, if someone or something is to be blamed, it must be the states where most of the responsible and capable people reside.

Edward Page points out that Simon Caney's cosmopolitan hybrid account of climate justice, which bases responsibility on both causal responsibility and ability to pay, locates the source of justice in the interests of persons.[4] But Page, whose own account of climate justice focuses on the rights and duties of states, criticizes Caney in part because it is not clear to Page how Caney's 'methodological individualism' can be 'operationalized given the national focus of the current global climate architecture, or in the face of widespread belief in the political and ethical sovereignty of individual countries'.[5] Page asks, for example, how Caney can help us to know 'which individuals in the developed world should contribute'.[6] Thus in Page's critique we see two problems with extant doctrine of climate justice: the obsession with sovereign states and the focus on people in developed countries only.

How can we address these two preoccupations? Caney's brand of cosmopolitanism can offer some answers. We need not abandon the state, but we should not fully accept the prevailing system either. Doing so in practice means muddling along while the Earth grows hotter and climate change does more harm and manifests itself in more human suffering. What is needed is an admission that the ethical sovereignty of states is not exclusive. A cosmopolitan corollary to the doctrine of international environmental justice could be a realistic and principled median between Page's (and other scholars') international climate justice, and some kind of global institution that, if ever created, will come about much too far in the future to help the world mitigate and cope with climate change in coming decades.

Using cosmopolitanism to help guide the world to better climate change policies requires us to focus much more attention on the world's polluters, including affluent new consumers in the developing world. This alteration to the usual cosmopolitan response is perhaps radical because cosmopolitans do not usually talk about duties in developing countries, but doing so is required by the rise of the new consumers. Furthermore, while practical cosmopolitans are right to recognize (if not endorse) the role of states, the orientation needs to change. Rather than advocating that states bear new burdens based on cosmopolitan morality – although they should do this – a more effective approach may be to benefit from cosmopolitanism's focus on the rights and duties of persons. States, rather than being the sole practical bearers or objects of cosmopolitan duties and rights, should instead be viewed more explicitly as facilitators of *individual* rights and duties. This approach may at first appear to be subtle, but its implications must be less so if the most disastrous consequences of climate change are to be avoided or significantly mitigated.

Putting cosmopolitanism into practice could mean a number of possible forms of governance and institutions. It could mean a world government, but it need not do so. As Patrick Hayden points out, 'cosmopolitanism is not inherently opposed to the state *per se* . . . Rather cosmopolitanism is generally concerned to develop varied modes of governance – from the local to the global – with the goal of facilitating the rights and interests of individuals *qua* human beings. Indeed, states may be one mode of governance well suited to this end. . .'.[7] David Schlosberg argues that 'institutions of engagement' to bring about environmental justice 'could not exist solely at the state level; the focus must be at multiple levels – including both the state political realm and the transnational level'.[8] While it might be ideal for the world to be governed by truly cosmopolitan institutions, they are unlikely anytime soon. As Jon Mandle puts it, 'the world today and for the foreseeable future is one in which individuals and corporate actors pursue their goals against a background of rules and institutions created by states'.[9] Axel Gosseries believes that adopting the assumptions of cosmopolitanism need not stop us from 'using states as our point of reference [because states can be] most able to represent the individuals that constitute them and because they are currently the most relevant units in the context of global attempts of curbing [greenhouse gas] emissions'.[10] Caney advocates a kind of 'revised statism' in which states do more to promote and implement global justice.[11]

Future institutions could be premised on the understanding that the 'central argument of contemporary cosmopolitan political thought is that the demands of justice must be decoupled, at least to some degree, from the territorial bounds of the states'.[12] Thomas Pogge argues that externalities like climate change 'bring into play the political human rights of . . . outsiders, thereby morally undermining the conventional insistence on an absolute right to national self-determination'.[13] He proposes a dispersion of state sovereignty premised on cosmopolitan morality.[14] For Lorraine Elliott, 'an ecologically sensitive cosmopolitanism demands transnational environmental justice between people within a world society as well as, and possibly in preference to, international justice between states in an international society'.[15] So we are left with something less than world government, and certainly a continuing role for states, but institutions informed by moral cosmopolitanism's advocacy of the equal worth of all persons and the need for protecting their basic rights. Consequently, following the moral cosmopolitan belief that 'every human being has a global stature as the ultimate unit of moral concern',[16] but without rejecting the state system generally and the climate change regime in particular, a cosmopolitan corollary to international doctrine would establish a much more prominent place for human beings – their rights, responsibilities and duties – in the evolving climate change regime.

Features of a Cosmopolitan Corollary

A cosmopolitan corollary to international doctrine would include several features. Andrew Linklater's description of a cosmopolitan, 'post-Westphalian' world is one based in part on the premise that 'the primary duty of protecting the vulnerable rests with the source of transnational harm and not with the national governments of the victims'.[17] This implies that the source of the harm, and the responsibility not to cause harm in the first place, rests with actors other than national governments, or at least in addition to them. From a cosmopolitan perspective, this responsibility rests with people regardless of the national governments ruling the place where they live. Cosmopolitan diplomacy surrounding climate change should be premised on the rights and duties of human beings. Persons should be at the centre of climate change discourse, negotiations and policies, and they should be the viewed as the primary *ends* of diplomacy and government policy, comparable to Kant's imperative to treat human beings as ends rather than means to an end (that is, protecting state interests). Thus, while governments will inevitably retain a central role, as even most cosmopolitans recognize (at least because states are unlikely to surrender their role), unlike even most cosmopolitan arguments (which value persons but translate that into interstate duties, usually for rich states only), a cosmopolitan corollary would have governments playing the role of facilitators of global citizenship and the implementation of cosmopolitan obligation.

The ethical advantages are clear from a cosmopolitan perspective – human beings move to the centre. But equally important is the practical impact of more consumers and polluters being brought into the discourse of, and the solutions to, climate change. Very crucially, there are political benefits as well: states are, to some extent, taken off the hook, greatly reducing their political reflex to resist genuine collective action on climate change, in part because their citizens will at least see everyone in the world being treated more or less equally based on their conditions in life.

Additional political advantages come from governments of the economic benefits that can accrue to many of their constituents from implementing a cosmopolitan corollary.

A cosmopolitan corollary to the doctrine of international environmental justice would start with recognition of global justice and the rights of all persons and the duties of capable persons, and includes conscious efforts to actualize global justice in agreements and the national and multinational institutions of states. A corollary comprises two fundamental changes to the manner in which climate change is dealt with at the diplomatic level: (1) a change in the official discourse so that it acknowledges and affirms the rights and duties of all people in the context of climate change, and (2) the incorporation of human rights and responsibilities into international agreements on climate change. From this new perspective, people become primary objects and explicit ends of political and diplomatic negotiations, agreements and policies. International agreements and resulting policies would aim to promote a global environment suitable for human well-being and flourishing. States become facilitators of human rights and protectors of human interests related to climate change – a 'responsibility to protect' people and their 'human sovereignty' (to invoke two important concepts increasingly accepted by the international community) from the adverse impacts of climate change, in keeping with the notion that states exist to protect people and their well-being.

A cosmopolitan corollary would mirror Hayden's suggestion that 'the discourse of sustainable *development* should give way to a discourse of sustainable or environmental *justice*'.[18] A corollary would be layered with existing national and international responses to climate change. That is, a global justice approach to addressing climate change is one that integrates discourse and thinking about, and action by, people as an add-on or supplement to more traditional communitarian approaches (without being subsidiary to them). As a corollary to the doctrine of international environmental justice, it is aimed at expanding the scope of climate justice by combining international justice with responses premised on world ethics. In short, human beings become a central moral basis of climate change diplomacy and policy, in effect making the climate change regime more ethically sound and more likely to elicit necessary action.

Elliott rightly suggests that we must doubt whether 'relying on states not just as moral agents but as the moral subjects . . . is sufficient to ensure that individuals and their communities will be treated justly'.[19] One might respond that we ought to do away with the role of states. But even if that were a long-term aim, in the meantime a first step ought to be to promote individuals as moral agents even as states remain largely in control. This could be the starting point for an alternative to the state-centric tragedy of the atmospheric commons. A cosmopolitan corollary would be, in essence, cosmopolitanism grafted on to extant Westphalian norms and the doctrines of international relations that have so far guided climate diplomacy and policy responses. In this respect it is a kind of bridge across the divide between the nation-state system and the imperative of climate protection. This process of making *both* persons (and their rights and responsibilities) and states (and their rights and responsibilities) objectives of the climate change regime is the central feature of a cosmopolitan corollary. By more explicitly encompassing both states and persons, it builds on and corrects the existing state-centred regime.

At first it may seem that a cosmopolitan corollary is nothing more than doing what many cosmopolitans already advocate: moral cosmopolitanism implemented by state institutions, whether they are at the national or international level. But there is a fundamental point that bears emphasis: rather than being about how states can, at best, implement cosmopolitan principles, a cosmopolitan corollary would be about how states can help *persons* implement cosmopolitan duties and enjoy related rights. As such, a cosmopolitan corollary would be about moral ethics and practical ethics for people, while recognizing that states cannot realistically be removed from the mix. This apparently subtle difference between what most cosmopolitans do (because they are more realistic than their critics assume) – accepting, perhaps grudgingly, states and their institutions – and what a cosmopolitan corollary is intended to do – to be closer to the cosmopolitan ideal of *individuals* being at the centre of policy or, put another way, to more fully view states as incidental to world ethics – is more important than it might seem. This is because a corollary has potential *practical and political* significance. If cosmopolitanism is used only to find new rights and duties for states, which it most often is used to do in the context of climate change, then it does little to help us escape the curse of Westphalia – the blame game among states of 'you go first' that currently plagues the climate change negotiations – and might even make things worse by *reinforcing* the obligations of certain states over others, or at the very least affirming those obligations in the minds of the people that matter, such as government leaders and diplomats. Alternatively, a cosmopolitan corollary should have the effect of helping states to free themselves from the myopia of states' rights and obligations by focusing on how practical progress on climate change can be made through recognizing and trying to actualize the rights and obligations of *affluent, capable persons everywhere*.

A cosmopolitan corollary to the doctrine of international environmental justice would be principled, practical and politically viable – probably even politically essential. Most of the ethical-normative arguments for cosmopolitanism and world ethics can be brought to bear to justify bringing global justice into the climate change debate. A cosmopolitan corollary to the doctrine of international environmental justice would therefore be principled; it would be the right thing to do for a number of reasons (for example, causality, capability, vital interests, human rights, harm and so forth). Perhaps most simply, a cosmopolitan corollary would be more principled and just because it attaches duties to those who cause global warming and advances the rights of those who suffer the most from it, regardless of their nationality. A cosmopolitan corollary would stop ignoring humans, their rights and their duties. Instead it would aim to recognize and promote the rights of people, especially the least well off, while incorporating the duties of all capable people, notably affluent ones all over the world, into efforts to address climate change. To be sure, for a cosmopolitan corollary to work – for it to contribute to ending the tragedy of the atmospheric commons that is the reality of a climate change regime premised on international doctrine – the principle of it would have to be taken seriously. This is not so much because principle matters, which of course it does for all sorts of reasons, but because if diplomats see cosmopolitanism as just another instrument for promoting state interests, they and their governments will fall back into the same tragic behaviour. By aiming the climate change regime at promoting cosmopolitan principles, which means putting people at the centre of the regime, diplomats can direct more attention to the causes and consequences of global warming and less to how traditional state-centric policies present problems for their states' perceived interests and long-held positions in the international relations of climate change.

A cosmopolitan corollary to the doctrine of international environmental justice would also be practical. It would reflect climate change realities rather than assuming that the problem, and all of the solutions to it, must or even can comport with the Westphalian assumptions of state sovereignty, rights, autonomy and independence. Unlike current doctrine, it would focus on the actual source of much of the world's greenhouse gas pollution – individuals. It would directly address the increasingly important role played by large numbers of newly affluent people in the developing world while still fully encompassing affluent people in the industrialized world and while recognizing that many poorer people in the latter are relatively minor contributors to the problem or not really capable of taking on obligations related to climate change. Pogge has argued that people in wealthy countries should not avoid taking responsibility for the harm they cause to people in poor countries simply because *some* of the latter 'will get away with murder or with enriching themselves by starving the poor . . . This sad fact neither permits us (affluent people in affluent countries) to join their ranks, nor forbids us to reduce such crimes as far as we can.'[20] We must agree with Pogge *and* take his argument further: we can no longer ignore affluent people in poor countries because it is no longer *some* of them committing 'murder', but now many millions of them. Put another way, the duty of most people in developed countries to act is undiminished, but we must stop letting all of the new consumers, and even rich elites in developing countries, get away with murder as their counterparts in the West have done for so long. Even if we accept that cosmopolitanism, world ethics and global justice are idealistic in other spheres of human activity, in the case of climate change they are utterly practical and necessary. The degree to which a cosmopolitan corollary would be practical depends in large measure on how it is conceived and implemented. But this is just as true of other ideas for addressing climate change.

A cosmopolitan corollary to existing interstate doctrine underlying the climate change regime is also *politically viable* and likely to be politically essential if the world is to salvage the regime and move it toward much more robust outcomes, especially in terms of mitigation but also in terms of adaptation and even compensation. It is perhaps important to reiterate at this point that what is being proposed is not to replace the doctrine of international environmental justice, but rather to supplement it and build upon it: to formulate and to implement a corollary to the international environmental justice doctrine. Thus a corollary is unlikely to face the kind of opposition from people and governments that would be experienced by wholesale change, and perhaps it would be embraced. What is more, if the world's new consumers are brought into the climate change regime, as a corollary would require, people in rich countries will see affluent people in poor countries responding. This would make it far easier for governments of the developed countries to sell the climate change regime to their citizens. At the same time, governments of developing countries can agree to limit the greenhouse pollution of their affluent citizens without undermining longstanding demands for international justice. They need not do what they insist that they will not do: take on mandatory *national* commitments to cut emissions of greenhouse gases. They can tell their citizens that they have won the argument in this respect. But the result on the ground is that large numbers of people (albeit affluent minorities) in developing countries will start to limit their emissions even as majorities of people there (the poor) are not required to do and instead are aided to improve their living standards. Put another way, pollution among some groups *within* developing states will decline even though pollution *of* those states is not required to do so.

Additionally, insofar as affluent people everywhere contribute to global funds for dealing with climate change, the net effect will be more aid for people in developing countries (see below). *Everyone* could be equally obligated to pay in based on a 'cosmopolitan formula' that takes into consideration needs, capabilities, affluence, responsibility (and perhaps enjoyment of the fruits of past emissions) and so forth. While some funds would go to the poor in affluent countries, the net result is more funds going to help people in developing countries, making a new, more cosmopolitan regime much easier to sell there. This can be achieved without poor countries having to take on any new (and unfair) international burdens, yet they (actually, only some of their citizens) will be seen by rich states (and their citizens) to be doing so by paying into a new climate fund and by limiting, if not always reducing, their greenhouse gas emissions.

A cosmopolitan corollary would not be politically idealistic, least of all utopian. The beauty of this approach is in part its political palatability. It gives states and diplomats the political cover they need by allowing them to stick with their longstanding principles. States *qua* states take on few new obligations when agreeing to a cosmopolitan corollary. Diplomats from developing states can go home and say, in all truthfulness, that they have not compromised their demands for international justice, and even that people of the developed countries have, through their governments, agreed to take on new commitments to cut greenhouse gas emissions. Diplomats from developed states can go home and claim, accurately, that people in developing countries have finally agreed, via their own governments, to take on new commitments to cut their greenhouse gas pollution. This gives an important political concession to the developing countries – they get respected and even paid as demanded and as is right, from the perspective of *international* justice – and it gives the developed countries what they want and that which is required – involvement of the developing world in emissions cuts. The psychological impact of this new paradigm on populations in Australia, Canada, the United States and other developed countries where people (and governments) have been waiting for developing countries to commit to concrete greenhouse gas limitations could be very powerful. It gives governments the political insulation they need to finally do what they know, as states, that they ought to have been doing for some time.

IMPLEMENTING A COSMOPOLITAN COROLLARY AMONG STATES

At the international level, a cosmopolitan corollary would involve changes in diplomatic discourse; changes in international agreements that explicitly invoke, recognize and incorporate cosmopolitan rights and duties;

changes in how those agreements are implemented by international institutions so that individuals' rights and duties are actualized; new kinds of funding mechanisms to distribute aid; and, ideally, representation of people as well as governments in negotiations. Even though the act of climate diplomacy is not cosmopolitan per se, comprising as it does relations among sovereign states and their representatives, implementing cosmopolitan imperatives should be one of its primary aims – not its only aim, not an aim to replace current aims premised on the seemingly inevitable baggage of extant international norms, not a utopian aim – but a practical aim based on the realities of climate change, its causes and its consequences. If states will not get out of the way (which they will not, realistically), they ought to at least become mediators of cosmopolitan duties; states and their international organizations ought to enable global citizenship, at least in the context of climate change, alongside national citizenship.

A key to a corollary is that requirements of global justice, such as recognition of human rights, distributive equity and increasing the capabilities of the disadvantaged, ought to be incorporated into agreements. What is more, procedural fairness, whereby *people* and their interests are part of the process of negotiating the agreements, should be implemented. According to a cosmopolitan corollary, the poor and underrepresented should not be merely a means to an end; they ought to be involved in the dialogue (as cosmopolitans often argue) and the *ends* of the climate change regime, in addition to ends focusing on states and other actors. It is important not only to empower poor states in climate change negotiations, but also to empower poor people, and indeed all people, or at the very least to try to represent their concerns, interests and even their aspirations. As a general rule, people significantly affected by political decisions and institutions have a right to be represented in making those decisions and running those institutions.[21] Importantly, diplomats and others should 'not look for a perfect system of representation before acting on the already obvious imperfect and biased system we have, and to bring a form of presence to those regularly left out of the decision-making process'.[22] Procedural fairness could involve proxies that might include representatives of various peoples, nongovernmental actors in some cases, and even existing governments that have been seated through fully democratic means. One example of involving people more in negotiations about climate change can be found in the Aarhus Convention on Access to Information, Public Participation in Decision-Making and Access to Justice in Environmental Matters of the European Region.[23] The Aarhus Convention provides legal remedies to enable citizens to challenge governments' denial of information and to increase participation by the public in international forums.[24]

One important example of how a cosmopolitan corollary might be implemented among states is new (or reformed) funding mechanisms. While funding is not the only method for implementing a cosmopolitan corollary, it is worth exploring here as an illustration of what a corollary might look like in practice. Additionally, funding is important because it is related to other aspects of dealing with climate change: funds can enable greenhouse gas emissions limitations, and they can be used to help people and communities adapt to climate change and also to compensate people harmed by it. The act of gathering funds can also influence behaviours (for example, taxes on fossil fuels can discourage their use). Coming up with substantial funds to, for example, assist those harmed by climate change need not be at all onerous. By way of illustration, Pogge points out that a 'global resource dividend' (that is, global tax) of only 1 percent of global product would raise several hundred billion dollars. Importantly, his global resource dividend and taxes on excessive energy use for climate adaptation and compensation are not a form of aid: 'It does not take away some of what belongs to the affluent. Rather, it modifies conventional property rights so as to give legal effect to an inalienable moral right of the poor.'[25]

Many cosmopolitans argue in favour of different kinds of global funds, with proceeds dispersed so that everyone has enough resources for a dignified basic existence, for the protection of human rights or for 'unconditional basic income', particularly when basic needs are denied as a consequence of global structures, such as the world economic order.[26] Brian Barry suggests that cosmopolitanism is 'best satisfied in a world in which rich people wherever they lived would be taxed for the benefit of poor people wherever they live',[27] thereby reducing the role of sovereign states while allowing them a role for raising funds and allowing international organizations a role for distributing those funds.[28] In principle, cosmopolitanism would suggest that global distribution of income be derived from an income tax 'levied at the same rate on people with the same income, regardless of where they live', and those who receive the resulting funds 'should be poor people again regardless of their place of residence'.[29] This could be the general model for a climate change-related tax premised on global justice. Among specific measures could be a carbon tax on greenhouse gas emissions, which Barry says would ideally be collected 'directly from the users or polluters', which is preferable to taxing states based on per capita national incomes because 'individual income acts as a proxy for resource use wherever the person with income lives'.[30] More of the money should come from earmarked climate change-related taxes on non-essential activities. This would include, among other things, taxing international airline flights, luxury goods and other non-essential polluting activities and goods.[31] This would raise new money and restrain harmful activity. Here we see the affluent aiding and acting to address climate change, in congruence with aims of cosmopolitan justice.

Peter Barnes has proposed a 'sky trust' that would administer greenhouse gas emissions rights on behalf of the world's citizens and pay out dividends to everyone based on its income.[32] This could work by charging oil and coal companies for emissions rights, which would raise the cost of carbon intensive energy and require people who use it to pay higher prices, while also generating fees for the trust. This would reduce demand and generate income that could be distributed to everyone, meaning that those who use the least of the polluting energy sources (that is, the world's poorest people) would benefit the most. The fact that the funds would be redistributed to everyone, including the poor who need the higher-priced fuels, 'would help to gain public recognition for the idea that, despite different emission levels, all citizens have an equal *per capita* right to the atmosphere'.[33] Hillel Steiner describes a global fund, to which each state has an equal per capita claim, that is derived from aggregating the individual claims of people in that state.[34] He says that 'each person – regardless of where on the globe he or she resides – is owed that equal amount', the payment of which could take the form of 'unconditional basic income' or an 'initial capital stake' or some other *equal per capita* form.[35] Steiner believes that such a fund 'would serve to establish a variety of benign incentive structures informing relations both within and among nations'.[36]

Jouni Paavola argues for institutionalizing responsibility for greenhouse gas emissions through a uniform carbon tax implemented at the national level.[37] He recommends a low tax-free quota, which would put the bulk of the tax burden on already developed states, but he also advocates extending this tax to other countries 'when they become significant per capita greenhouse gas emitters'.[38] This latter point raises the question of what happens if per capita emissions in developing countries go over the allowable global per capita level, as many will, even while those countries remain poor. The answer should be to go ahead and implement the tax, qualifying it for poor people who have no choice but to exceed the acceptable level, but of course making sure that it captures the truly affluent in those poor states. As Paavola suggests, his tax proposal would create incentives for efficient energy choices, and the resulting 'combined compensation and assistance fund' would provide revenue to be used 'for compensating the impacts of climate and for assisting adaptation to climate change'.[39] Indeed, if poor people and communities in *all* countries could draw on the fund – that is, including people in developed countries – it could provide important incentives for their governments, most importantly, developed country governments, to take on their responsibilities, and would certainly mitigate the disincentives they have for not doing so up to now.

The United Nations could administer funding to limit climate change and aid those who suffer from it the most. Some or much of the money raised from climate-related taxes could be deposited into existing funds, such as the Global Environment Facility, the Special Climate Change Fund, the Least Developed Country Fund and the Kyoto Protocol Adaptation Fund. These funding mechanisms are hardly ideal, but they are increasingly the kind of thing that is needed. There might also be a new fund, perhaps a Future Climate Fund, specifically designed to aid future generations, possibly funded primarily from a tax on fossil fuels used by affluent people everywhere, to help future generations cope with climate change caused by past, present and future greenhouse gas emissions. A new climate-fund scheme would have an important difference compared to existing schemes: contributions to it would be based on individuals responsibilities, duties and capabilities, and payouts, while administered by international organizations and capable and willing state governments (or perhaps nongovernmental organizations when states are unwilling or unable to administer the funds, as with some development and disaster aid today), would be based on individual responsibilities, duties and capabilities. This would mean, for example, that a New York City executive would pay into the fund, but so too would a Hong Kong executive. In aggregate, of course, all affluent Americans (that is, the United States and most of its citizens) would pay far more into the fund than they would receive, and all Chinese (*qua* China) would receive far more than China's affluent citizens pay into the fund. But rich Chinese would no longer be treated just like poor ones – and very poor Americans would no longer be treated as though they are rich. Individuals can and should also give money to nongovernmental organizations doing credible work to alleviate the suffering of those affected by climate change now and in the future.

Some countries in Europe have started to levy taxes and auction pollution rights, with some of the resulting funds going to climate adaptation programmes in developing countries. Twenty per cent of funds collected from the European Union's Emissions Trading Scheme (about US$2 billion per year by 2020) are to be allocated to climate change-related projects, including adaptation in developing countries.[40] One feature of the 2007 Bali Roadmap was a protocol for collecting a levy on Clean Development Mechanism projects, the funds from which are to be contributed to the Kyoto Protocol's adaptation fund. In 2008 governments agreed that that the levy would amount to 2 percent of the value of carbon credits that developed countries derive from the mechanism's projects in developing countries, with predictions that resulting funds would be as much as US$950 million by 2012.[41] Alas, the amount of funding from this and other sources so far 'is just a puff of smoke' compared to the many tens of billions per year (at least) that poor countries and their people will need to adapt to climate change.[42]

The formula for who should pay into climate change funds, and the amount they should pay, could be based on equal per capita shares as a starting point, with other considerations being factored in (for example, responsibility, capability). The same kind of formula could apply to cuts, limits or increases in greenhouse gas emissions by and for individuals. (The majority of the world's people ought to be allowed and even empowered to increase their emissions while the world transitions to a post-carbon energy system.) The funds resulting from a cosmopolitan corollary could pay for things like disaster relief, poverty alleviation, sustainable development, greenhouse gas mitigation, adaptation measures, technology transfers and the like – and even for compensation. The ultimate ends of this funding must be human beings, although one assumes that some of the money might go to programmes to improve countrywide economies that would benefit as many people as possible going forward. One aim of the fund should be to discourage growth in populations where people become affluent, at least until affluence and fossil-fuel use can be decoupled (which is likely to happen without much encouragement because the most developed places tend to have low population growth, or even population decline). Importantly, as people in poor countries become wealthier, most of the benefits that accrue to them from the climate fund would go down. The aim would be for them to one day become affluent enough to stop receiving aid and to start paying into the fund, freeing up more money for those most in need, creating a kind of virtuous and increasingly effective cycle of funding for climate-related objectives.

The incentives and disincentives resulting from collecting such funds, if based on climate-related criteria such as energy use (causal criteria) and poverty (consequential criteria), could be used to promote greenhouse gas mitigation and the redistribution of wealth across borders, mainly to the advantage of people in developing states – thus giving their governments incentives to participate – and to the least well-off people everywhere, including in developed states – thus also giving their governments some incentive to participate, or at least lowering opposition to doing so. While such funds would be *international*, they would be fundamentally structured on cosmopolitan principles – on per capita bases in terms of fundraising and payouts.[43] There would have to be incentives and schemes for states unwilling or unable to implement a corollary as manifested in new agreements, and money might have to be given to nongovernmental actors or international financial institutions to disperse (or to be put into escrow for when those actors are allowed by governments to aid people). Given that this responsibility to aid is largely (but not wholly) based on responsibility for suffering, affluent individuals in affluent states might have more obligation to provide aid because they often benefit more from their own and others' past pollution. However, the responsibility of affluent people in less affluent countries to aid starts from the moment they live an affluent life, and of course increases the more they consume and pollute – in addition to inherent obligations on all affluent and capable people, regardless of how much they pollute, to aid the poor. Importantly, none of this absolves affluent governments from continuing and increasing the types of international transfers that obtain at present or are already envisioned in the context of the climate change regime.

Actualizing these or other sorts of schemes for funding and otherwise acting on a cosmopolitan corollary among states would admittedly run up against many of practical obstacles, but Barry confronts this head on: 'unless the moral case is made, we can be sure nothing good will happen. The more the case is made, the better the chance.'[44]

CONCLUSION

At least in the context of climate change, cosmopolitanism – and specifically a cosmopolitan corollary to the doctrine of international environmental justice – is principled, practical and politically viable. This contrasts with the purely interstate approach that has generally guided the climate change regime. That approach has been ethically deficient, ignoring as it does the rights and obligations of people, and it is very limited in its practicality because it is premised on narrow state interests and thus has resulted in little collective action to combat climate change seriously, in the process stifling politically innovative, robust and truly effective solutions to climate change. As Elliott notes, 'the normative interests of the state remain evident in the dominance of sovereignty claims and national interest that are pursued at the expense of cosmopolitan values and at the expense of the environment. The state therefore remains ambiguous as cosmopolitan moral agent.'[45] Even as we look to states to assist in fostering global environmental justice, it is worrying that they have utterly failed to arrest global warming and respond to climate change effectively, let alone done so in a way that promotes justice, whether among states or among people. Consequently, while there is too little time to replace states, meaning that we must aim to reform their practices through injections of cosmopolitanism, the fundamental truth is that most of the important solutions to climate change are in the hands of people.

To suggest that persons ought to become much more of a subject of diplomacy and policy among states is not radical. Indeed, as Christian Reus-Smit has pointed out, one important feature of today's international system is a:

> progressive 'cosmopolitanisation' of international law, the movement away from a legal system in which states are the sole legal subjects, and in which the domestic is tightly quarantined from the international, toward a transnational legal order that grants legal rights and agency to individuals and erodes the traditional boundary between inside and outside.[46]

One assumes that one thing driving this evolution in international law is a growing sense that persons per se have moral standing. This shift offers a good basis for a cosmopolitan corollary to international doctrine generally and international environmental justice in particular.

Nigel Dower identifies a key question that ought to guide climate diplomacy and policy: 'a cosmopolitan view will require us to look very hard at policies with a view to answering the question: does this contribute to or avoid not impeding the overall global good *vis-à-vis* the environment?'[47] To quote Onora O'Neill, 'any theory of justice that wishes to be taken seriously must respect empirical findings'.[48] The doctrine of *international* justice and its underlying normative foundations have failed these tests. The statist doctrine has failed to give sufficient moral and practical weight – and usually fails even to acknowledge – the empirical reality of the shifting balance of greenhouse gas emissions away from being mostly a collective act of people in developed countries, which might have once made the climate change regime's focus on states practically reasonable, to one in which the emissions of affluent people in developing countries are rapidly rising toward those of people in developed countries. Thus an approach to justice that respects reality is one that at least incorporates, but ideally embraces, cosmopolitan ethics because it is that source of justice – it is *global* justice – which captures the full moral, practical and political facts of climate change in a globalized world.

A cosmopolitan corollary to international justice offers an escape from the legal and mental straightjacket of Westphalian norms. By associating the pollution of individuals and classes of people with ethical diplomatic arguments, international agreements and the domestic policies intended for implementation of those agreements, governments of both developed and developing states can escape the ongoing blame game in which poor states blame rich ones for the problem's creation, and rich states blame poor ones for the problem's future trajectory – with both refusing to sufficiently obligate even their affluent citizens to do all that is necessary and just. In the context of climate change, cosmopolitan justice has the potential to define a pathway whereby major developing country governments can simultaneously assert and defend their well-justified arguments rejecting *national* climate change-related obligations while also acknowledging and regulating growing pollution among significant segments of their populations. This in turn can help to neutralize the reticence of most developed country governments and their publics to live up to their states' obligations to finally undertake the major cuts in greenhouse gas emissions that will be required to limit future damage to the atmospheric commons upon which all states and all people depend. A cosmopolitan corollary can also help to free up new financial resources to aid those people most harmed by climate change. The conclusion we are left with is that cosmopolitan (or global) justice is almost certainly unavoidable if climate change is to be addressed effectively anytime soon.

NOTES

1. This chapter summarizes parts of Paul G. Harris, *World Ethics and Climate Change: From International to Global Justice* (Edinburgh: Edinburgh University Press, 2010), especially Chapter 7, 'Cosmopolitan diplomacy and climate policy'; and Paul G. Harris, 'Reconceptualizing governance', in *Oxford Handbook of Climate Change and Society*, John Dryzek, Richard Norgaard and David Schlosberg (eds) (Oxford: Oxford University Press, 2011), pp. 639–652.
2. James Gustave Speth, *The Bridge at the Edge of the World* (London: Yale University Press, 2008), p. 29.
3. See Norman Myers and Jennifer Kent, *The New Consumers* (London: Island Press, 2004).
4. Edward A. Page, 'Distributing the burdens of climate change', *Environmental Politics* 17, no. 4 (2008): 570. See, for example, Simon Caney, 'Cosmopolitanism, democracy and distributive justice,' in *Global Justice, Global Institutions*, Daniel Weinstock (ed.) (Calgary: University of Calgary Press, 2007), pp. 29–64.
5. Page (2008), p. 570.
6. *Ibid.*
7. Patrick Hayden, *Cosmopolitan Global Politics* (Aldershot: Ashgate, 2005), p. 21.
8. David Schlosberg, *Defining Environmental Justice* (Oxford: Oxford University Press, 2007), p. 188.
9. Jon Mandle, *Global Justice* (Cambridge: Polity, 2006), p. x.
10. Axel Gosseries, 'Cosmopolitan luck egalitarianism and the greenhouse effect', in *Global Justice, Global Institutions*, Daniel Weinstock (ed.) (Calgary: University of Calgary Press, 2007), p. 280.
11. Caney (2007).
12. Aaron Maltais, 'Global warming and the cosmopolitan political conception of justice', *Environmental Politics* 17, no. 4 (2008): 594.
13. Thomas W. Pogge, *World Poverty and Human Rights*, 2nd ed. (Cambridge: Polity, 2008), p. 192.
14. *Ibid.*, p. 184.

15. Lorraine Elliott, 'Transnational environmental harm, inequity and the cosmopolitan response', in *Handbook of Global Environmental Politics*, Peter Dauverge (ed.) Cheltenham: Edward Elgar, 2005), p. 497.
16. Thomas Pogge, *World Poverty and Human Rights* (Cambridge: Polity, 2002), p. 169.
17. Andrew Linklater, *The Transformation of Political Community* (Cambridge: Polity, 1998), p. 84, quoted in Andrew Dobson, 'Globalisation, cosmopolitanism and the environment', *International Relations* 19, no. 3 (2005): 268–269.
18. Hayden (2005), p. 132.
19. Lorraine Elliott, 'Cosmopolitan environmental harm conventions,' *Global Society* 20, no. 3 (July 2006): 363.
20. Thomas Pogge, 'Real world justice', *Journal of Ethics* 9 (2005): 53.
21. See Pogge (2002), p. 184.
22. David Schlosberg, *Defining Environmental Justice* (Oxford: Oxford University Press, 2007), pp. 195–196.
23. See Wolfgang Sachs (ed.) *The Jo'burg Memo* (Berlin: Heinrich Boll Foundation, 2002), p. 52.
24. Jonas Ebbesson, 'Public participation', in *The Oxford Handbook of International Environmental Law*, Daniel Bodansky, Jutta Brunnee and Ellen Hey (eds) (Oxford: Oxford University Press, 2007), pp. 692, 701.
25. Pogge (2005), p. 52.
26. Gillian Brock and Harry Brighouse, 'Introduction', in *The Political Philosophy of Cosmopolitanism*, Gillian Brock and Harry Brighouse (eds) (Cambridge: Cambridge University Press, 2005), p. 8.
27. Brian Barry, *Justice as Impartiality* (Oxford: Clarendon Press, 1995), p. 153.
28. To avoid the familiar problem of the rich in poor countries stealing the funds, the transfers might have to be made directly to individuals rather than via governments.
29. John Barry, 'Statism and nationalism: a cosmopolitan critique', in *Global Justice*, Ian Shapiro and Lea Brilmayer (eds) (New York: New York University Press, 1999), p. 40.
30. Brian Barry, 'International society from a cosmopolitan perspective', *International Society* (Princeton: Princeton University Press, 1998), p. 155.
31. As always, the super rich will simply pay taxes on activities that are not regulated.
32. Peter Barnes, *Who Owns the Sky?* (Washington: Island Press, 2000).
33. Wolfgang Sachs and Tilman Santarius, *Fair Future* (London: Zed, 2007), p. 190.
34. Hillel Steiner, 'Territorial justice and global redistribution', in *The Political Philosophy of Cosmopolitanism*, Gillian Brock and Harry Brighouse (eds) (Cambridge: Cambridge University Press, 2008), p. 36.
35. *Ibid.*
36. *Ibid.*
37. Jouni Paavola, 'Seeking justice: international environmental governance and climate change', *Globalizations* 2, no. 3 (2005): 317.
38. *Ibid.*
39. *Ibid.*
40. The Economist, 'Climate change and the poor: adapt or die', *Economist*, 11 September, 2008.
41. *Ibid.*
42. The United Nations Development Programme, Human Development Report 2007/2008 (New York: United Nations, 2007), p. 194, gives a 'lower ballpark' figure of US$86 billion annually for costs of adaptation to climate change in 2015, while Christian Aid places costs at US$100 billion per year (Karrine Haegstad Flam and Jon Birger Skjaerseth, 'Does adequate financing exist for adaptation in developing countries?', *Climate Policy* 9 (2009): 110).
43. See Paul G. Harris and Jonathan Symons, 'Justice in adaptation to climate change: cosmopolitan implications for institutions', *Environmental Politics*, 19, no. 4 (2010): 617–636.
44. Barry (1998), p. 156.

45. Elliott (2005), p. 498.
46. Christian Reus-Smit, 'Introduction', in *The Politics of International Law*, Christian Reus-Smit (ed.) (Cambridge: Cambridge University Press, 2004), p. 7.
47. Nigel Dower, *World Ethics*, 2nd ed. (Edinburgh: Edinburgh University Press, 2007), p. 186.
48. Onora O'Neill, *Bounds of Justice* (Cambridge: Cambridge University Press, 2000), p. 2.

Index